D1261596

Public Lands
Conflict and Resolution

Managing National Forest Disputes

ENVIRONMENT, DEVELOPMENT, AND PUBLIC POLICY

A series of volumes under the general editorship of
Lawrence Susskind, *Massachusetts Institute of Technology*

ENVIRONMENTAL POLICY AND PLANNING

Series Editor:
Lawrence Susskind, *Massachusetts Institute of Technology, Cambridge, Massachusetts*

BEYOND THE NEIGHBORHOOD UNIT
Residential Environments and Public Policy
Tribid Banerjee and William C. Baer

ENVIRONMENTAL DISPUTE RESOLUTION
Lawrence S. Bacow and Michael Wheeler

ENVIRONMENTAL LAW AND AMERICAN BUSINESS
Dilemmas of Compliance
Joseph F. DiMento

THE LAND USE POLICY DEBATE IN THE UNITED STATES
Edited by Judith I. de Neufville

PATERNALISM, CONFLICT, AND COPRODUCTION
Learning from Citizen Action and Citizen Participation in
Western Europe
Lawrence Susskind and Michael Elliott

PUBLIC LANDS CONFLICT AND RESOLUTION
Managing National Forest Disputes
Julia M. Wondolleck

RESOLVING DEVELOPMENT DISPUTES THROUGH
NEGOTIATIONS
Timothy J. Sullivan

Other subseries:

CITIES AND DEVELOPMENT
Series Editor:
Lloyd Rodwin, *Massachusetts Institute of Technology, Cambridge, Massachusetts*

PUBLIC POLICY AND SOCIAL SERVICES
Series Editor:
Gary Marx, *Massachusetts Institute of Technology, Cambridge, Massachusetts*

Public Lands
Conflict and Resolution
Managing National Forest Disputes

Julia M. Wondolleck

School of Natural Resources
The University of Michigan
Ann Arbor, Michigan

Plenum Press • New York and London

Library of Congress Cataloging in Publication Data

Wondolleck, Julia Marie.
   Public lands conflict and resolution: managing national forest disputes / Julia M.
Wondolleck.
      p.   cm.—(Environment, development, and public policy. Environmental
policy and planning)
   Bibliography: p.
   Includes index.
   ISBN 0-306-42861-X
   1. United States—Public lands. 2. Forest reserves—United States. 3. Forest policy—
United States. I. Title. II. Series.
HD205.W66   1988   88-13991
333.1′0973—dc19   CIP

© 1988 Plenum Press, New York
A Division of Plenum Publishing Corporation
233 Spring Street, New York, N.Y. 10013

Printed in the United States of America

To Steve

Preface

The United States Forest Service, perhaps more than any other federal agency, has made great strides during the past two decades revolutionizing its public involvement efforts and reshaping its profile through the hiring of professionals in many disciplinary areas long absent in the agency. In fact, to a large extent, the agency has been doing precisely what everyone has been clamoring for it to do: involving the public more in its decisions; hiring more wildlife biologists, recreation specialists, sociologists, planners, and individuals with "people skills"; and, furthermore, taking a more comprehensive and long-term view in planning the future of the national forests. The result has been significant—in some ways, monumental—changes in the agency and its land management practices. There are provisions for public input in almost all aspects of national forest management today. The professional disciplines represented throughout the agency's ranks are markedly more diverse than they have ever been. Moreover, no stone is left unturned in the agency's current forest-planning effort, undoubtedly the most comprehensive, interdisciplinary planning effort ever undertaken by a resource agency in the United States.

Regardless of the dramatic change that has occurred in the U.S. Forest Service since the early 1970s, the agency is still plagued by conflicts arising from dissatisfaction with how it is doing business. From timber, minerals- and fuels-industry interests to local, regional, and national environmental organizations, attention to and criticism of the agency's activities has been unrelenting. It is impossible to look at an agency estimating that it will receive upward of one thousand administrative appeals on its forthcoming forest plans and not wonder if something is not amiss.

The U.S. Forest Service has found itself between a rock and a hard place. It is in the unenviable position of choosing between legitimate and reasonable claims that, unfortunately, compete for the same limited resources. Whichever way the agency goes, it only lands right in the middle of a battlefield with its hands tied by administrative appeals, lawsuits, or Congressional intervention. Forest Service officials are often frustrated by the seeming futility of their land management efforts. Environmental and industry groups and local communities are similarly frustrated by what they perceive to be an unresponsive bureaucracy. There must be a better way.

The purpose of this book is to try to help define that "better way." In this respect, the story told here should be a promising one. Whereas many chapters do document and lament the agency's current malaise and much that is seemingly "wrong" in current forest management decision-making procedures, the intent is solely to pinpoint where the heart of this pervasive conflict lies in order to understand how we might better manage it. Although many have asked the question, *Who* might better manage the national forests given the current impasse?, I take a different approach and ask, *How* might the U.S. Forest Service better fulfill its complex mandate? *How* might the agency provide opportunities in its decision-making processes that meaningfully involve affected national-forest-user groups in representing their own values and concerns and, in so doing, satisfy them that these values and concerns are indeed being accommodated? *How* and when might the agency provide incentives for collaboration among its constituent groups rather than the now all-too-predictable adversarial battling? And, perhaps most critically, *How* might the agency rebuild public trust in it?

Since the mid-1970s, there has been considerable interest in and experimentation with decision-making processes specifically designed to resolve multiparty, public disputes, much as those plaguing national forest management. There are many people in the Forest Service and its constituent groups who are skeptical of the notion that some national forest management disputes might be resolved. They question how there can ever be a common ground between such diverse and combative interests as those making themselves a part of national forest management. Yet the idea is more than simply a notion, and that is what makes it so exciting. The idea is currently in practice, formally in some agencies and informally in the Forest Service itself. Given the historical willingness of the Forest Service to experiment, to be on the forefront of public involvement efforts at the federal level, and given how much it and its constituent groups potentially have to gain, the promise of such dispute resolution efforts is great. Moreover, as the cases described in this book illustrate, environmental and industry groups want to participate more directly in what happens in the national forest system because they have such a great stake in it. They want and need to know and understand the realistic constraints bounding the agency's actions as well as be a part of creative solutions that they can support and have some ownership in. Certainly, not all public land disputes are resolvable collaboratively, but it appears that some are if simply given the opportunity.

The dispute resolution ideas in this book are those of many practitioners and scholars in the field; I am truly indebted to them for both directly and indirectly sharing their experiences and enthusiasm with me. I want especially to thank Lawrence Susskind for his many insights in all phases of work on this book and for sharing his experiences and ideas on public dispute resolution and Lawrence Bacow who provided much patience, encouragement, and guidance on an earlier version of this book. I am also grateful to James Crowfoot for his continued support as I completed this book and for his fresh perspectives on my ideas. I also want to acknowledge the faculty, staff, and graduate students associated with the Environmental Impact Assessment Project at the Massachusetts Institute of Technology and the Environmental Conflict Project at the University of Michigan's School of Natural Resources.

Throughout my research, many people in the U.S. Forest Service, oil, gas, and timber industries, and environmental organizations gave generously of their time in helping me understand the specific events that transpired in several conflicts analyzed in this book. I would especially like to thank Al Reuter and Gary Marple of the Bridger-Teton National Forest; Phil Hocker of the Sierra Club; Kea Bardeen of the Mountain States Legal Foundation; Dick Hamilton of Getty Oil Company; and Story Clark of the Jackson Hole Alliance for Responsible Planning.

My gratitude goes as well to Matt Cullen and the Lincoln Institute of Land Policy for their financial support and much needed independence during the early part of this research; and to John F. Hennessey for his fellowship support during the first stage of this research at MIT.

I would like to thank Alan Altshuler for his valuable comments and suggestions on an earlier version of this book. I am also grateful to James Burchfield, Lisa Bardwell, Martha Tableman and Nancy Manring for their comments on different versions of the manuscript as it evolved and to Susan Yonts-Shepard for sharing her experiences and reflections on the potentials and problems of applying dispute resolution concepts and processes within the U.S. Forest Service.

Finally, words cannot express the debt of gratitude that I owe my parents, my family, and my friends for their support and encouragement throughout my work on this book. To Steve Yaffee, whose consistent encouragement, prodding, and understanding combined with invaluable scholarly guidance, has been the one constant throughout all stages of this book. And to Anna, whose little smile provided the sunshine during the final stages of this book and gave me the greatest incentive to finish it quickly.

Julia M. Wondolleck

Ann Arbor, Michigan

Contents

CHAPTER 1

Conflict and the Public Lands

When President Ronald Reagan appointed James Watt as his Secretary of the Interior, public lands issues were immediately thrust into the limelight. Watt's proposals were headline news in the western dailies, where public lands issues customarily predominate, as well as in the *New York Times* and other major eastern newspapers usually little-concerned with this topic. Watt's vision of the proper use and disposal of public lands was in direct contrast to that of his immediate predecessors and, as such, seemed a startling transformation of public land policy. Watt's positions were hardly unusual, however. His western development policies and the demands of the "Sagebrush Rebels" who supported him, when combined with the counterclaims of conservationists and preservationists, mirror much of public lands history. Decisions governing how the public lands should be used have always been controversial. In fact, the conflict generated by these decisions has been so intense that at times the history of public-land law development reads like the same action-packed Wild West stories that it inspired.

That public-land management decisions generated conflict in the past and continue to do so today should come as no surprise. These decisions allocate the tremendous wealth of resources contained in the public lands. With so much at stake, it should only be expected that the many groups affected by land management decisions will actively pursue decisions favorable to their interests. It should come as no surprise that their clamor reverberates in public-land agency offices in Wash-

1

ington and the field as well as in the halls of Congress. Nor should it be surprising that such long-prevailing disputes between different public-land-user groups only heighten as critical decisions draw near.

It is not this conflict *per se* that is of concern to public land management today. Conflict in and of itself is not inherently bad. In fact, sometimes it is good: it keeps federal officials alert, helps define issues, promotes checks and balances in agency decision-making, encourages creative solutions to problems, and ensures that the many interests at stake will be heard (Coser, 1956). What should be of concern is the outcome of this conflict: the apparent inability of administrative decision-making processes to resolve these inevitable disputes and therefore make decisions that are viable. Administrative processes that are not decisive are processes ill-suited to the problems they are meant to address.

This outcome sets today's conflict far apart from that of the past. Today, the process by which decisions are made is failing; decisions, once made, are unable to withstand the inevitable attacks of dissatisfied interest groups. Decisions are consistently being undermined through administrative appeals, lawsuits, and congressional action. The policies of Secretaries of the Interior and other federal officials, no matter how popular or despised, seem to have little effect on the indeterminate nature of public lands decisions.

The argument made in this book is that the problem currently posed by public lands conflict results from inadequate decision-making processes. Current administrative processes conform to a public land management paradigm that was developed in a different time, responding to different problems and based on assumptions that do not always hold true today.

Many efforts at reforming our public-land management system have been both suggested and attempted. Most have failed, however, because the true nature of the public-land management problem is not well understood. Why are public land decisions so readily contested and so easily undone? Where and why does administrative decision-making in practice diverge from how in theory it should occur? What do these differences between theory and practice reveal about the nature of the problem posed by public land management? Where and why have previous reforms succeeded and failed? What might be done to improve decision-making processes in order to avoid the now costly and inevitable consequences of public lands disputes? By understanding the problem currently posed by public land management in these terms, the administrative decision-making processes can be revised accordingly.

PUBLIC LAND MANAGEMENT: THE PARADIGM AND ITS SHORTCOMINGS

Public lands conflict today is generated by a question of who gets what: Should land be allocated to wilderness or development, timber or recreation, light recreation or heavy recreation, minerals development or wildlife management? These are resource allocation questions that have very large distributional consequences. In contrast, at the turn of the twentieth century a very different question confronted policy-makers. At that time, the central question that dominated policy debates was not, Who gets what?, but, much more fundamentally, Who should decide who gets what?, and, moreover, What principles should they use in deciding? The answer was clear to the conservationists of the progressive era who argued for rational land management based in scientific expertise (Hays, 1959):

> Since resource matters [are] basically technical in nature . . . technicians, rather than legislators, should deal with them. Foresters should determine the desirable annual timber cut; hydraulic engineers should establish the feasible extent of multiple-purpose river development and the specific location of reservoirs; agronomists should decide which forage areas could remain open for grazing without undue damage to water supplies. (p. 3)

Conservationists did not ignore the political dimensions of land management. Instead they directly attacked the dire consequences of political land management and compared them with the perceived virtues of scientific land management (Hays, 1959):

> Conflicts between competing resource users, especially, should not be dealt with through the normal processes of politics. Pressure group action, logrolling in Congress, or partisan debate could not guarantee rational and scientific decisions. Amid such jockeying for advantage with the resulting compromise, concern for efficiency would disappear. (p. 3)

Professional land managers succeeded in their effort to supplant political decision-making with scientific decision-making. Their response to public land issues at the turn of the century and their activism, then as well as now, in promoting and maintaining rational, scientific management was grounded in a particular land-management paradigm that persists today.

A paradigm, generally speaking, is a system of professional standards, behavioral norms, and conceptual approaches to problem-solving that becomes adopted and applied by a community, a profession, or a society (Kuhn, 1970). A paradigm is much more than a tradition.

Individuals adhering to a particular paradigm do not behave or decide in a specific manner because "that is the way things have always been done." Rather, they behave that way because it is perceived to be and accepted as the most appropriate way to act. Therefore, the paradigm directs a consistent response to problems as they arise. Unfortunately, paradigms can sometimes limit vision beyond their conceptual bounds to alternative solutions that might better solve a problem.

The public-land management paradigm, as it is depicted here, is the model of who and how land management should occur. The land management paradigm is the model of how professionals make decisions using a rational, scientifically-based analytic process. All land management problems are addressed consistently with this paradigm; all administrative decision-making processes are structured to coincide with its principles.

In the early part of the twentieth century, professionally applied scientific expertise was continually called upon as the U.S. Forest Service's management task became more complex and as conflicts in land uses arose. As a result of this continued reliance on scientific expertise and professional analysis, the paradigm was strengthened and became more engrained in land management processes. Historian Bernard Frank (1955) describes the response of the first U.S. Forest Service Chief, Gifford Pinchot, to the increasing complexity of Forest Service land management:

> [Pinchot] was deeply concerned about the inadequate knowledge of the best means of protecting and improving the resources of the newly created national forests. To remedy this serious deficiency, he added scientifically trained men to his staff. As the job of handling the national forests became more complex, as more and more people began to use them, and as interests began to clash, the need for carefully recorded observations and systematic methods of evaluating them became more urgent. (pp. 79–80)

The response is the same today. As resource conflicts heighten and as land management tasks become more complex, professional expertise in various scientific disciplines is increasingly called upon for answers. Therefore, the paradigm remains intact even though many issues remain unresolved after proceeding through current administrative processes.

The conservationists' land-management paradigm has successfully prevailed throughout the twentieth century. Today, however, it frequently fails when controversial decisions must be made. No matter how professional, scientific, and thorough their analyses may be, the resultant decisions are not as viable as they were in the past.

WHY THE CONTROVERSY?

Analysts have offered several reasons why land management decisions prove so controversial. Three common explanations are: First, one
resource-user group has amassed enough power over an agency to control that agency and ensure that all decisions advance the group's best
interests (*the capture theory*) (Barney, 1974; Culhane, 1981; Foss, 1960;
Shepherd, 1975; and, more generally, Lowi, 1969; McConnell, 1966).
Second, the agency itself has amassed enough power to ignore all resource-user groups and make decisions that enhance the organization's
power, prestige, and managerial discretion (Frome, 1974a,b; Reich,
1962; Schiff, 1962). Third, the agency's hands have been tied by budgetary constraints or explicit Congressional mandates (U.S. General Accounting Office, 1978). In each case, groups perceiving a bias against
their best interests, as a result of these situations, are compelled to
organize and fight to obtain more favorable decision outcomes. Conflict
is the result.

There is invariably some truth to each of the above assertions, depending upon the specific case reviewed. However, these explanations
are generally rejected in this analysis as lacking sufficient evidence to
explain why decision-making is at an impasse, and why the decisions
rendered are frequently not viable. Rather than consciously making decisions that defy the public's interest because they are either captured by
a single group, are serving their own political ends, or are merely following legislative or budgetary marching orders, public land agency officials
are generally trying in good faith to make decisions that promote the
public's interest. Problems arise, however, because their image of the
public interest does not always coincide with that of other land users.
And, furthermore, recent developments in public land and natural resources law have changed the balance of power in public land management by legitimizing many land uses and, thereby, the claims of their
advocates. As a result, decisions that fail to address the interests of any
particular resource-user group are not viable decisions; they are successfully contested as soon as they are made. The land management paradigm, premised on rational, scientifically based resource conservation
and use, is not equally able to accommodate the more recent and highly
judgmental preservation and noncommercial objectives. The strength of
this paradigm is now being tested as these new considerations are being
interjected into land managers' analyses. The decision-making processes oriented toward maintaining long-term yields from the public

lands are not effective at determining appropriate levels and allocations of noncommercial resources, such as wilderness, wildlife, recreation, and scenic amenities.

The transition from public-land management policies geared toward late nineteenth and early twentieth century conditions to new policies adjusted to late twentieth century objectives calls for reform of existing administrative procedures in a manner that may not always conform to the public-land management paradigm described here. These reformed processes must be designed to recognize and take advantage of the current balance of power among competing resource user groups and the legitimate claims of each. Failure to resolve the inevitable differences between the many affected groups only guarantees that the current impasse in decision-making will persist.

THE CASE OF THE NATIONAL FORESTS

The phrase *public land management* is all-encompassing. Under its rubric falls watershed and wildlife management, timber harvesting, mineral leasing, wilderness preservation, recreation provision, and range management, to name only the obvious. Several different federal agencies have management responsibilities involving public lands, particularly the Bureau of Land Management (BLM) and the U.S. Forest Service (USFS). To study each issue area and each land management agency in the context of this analysis would be both time-consuming and unnecessary. Instead, three contrasting issue areas, all in the national forest system, oil and gas leasing and permitting, timber management and harvesting, and national forest planning under the National Forest Management Act, are analyzed in order to understand the problem currently posed by public land management. With this understanding, specific reforms of the administrative decision-making process can be developed and studied.

The USFS and its management of national forest lands has been chosen as the focus of this study for several reasons. First, the agency has long fostered and abided by the paradigm described above. In fact, the Forest Service and its forefathers originally devised the current paradigmatic approach to land management. Second, the Forest Service has recognized many of the problems that will be described in this book and, perhaps more than any other agency, has made a concerted effort to try to address and resolve them. Agency officials have stressed more open decision-making, improved communication with user groups, and, furthermore, have expanded the agency's interdisciplinary profile. The agency's new programs and policies invariably have been guided by this land management paradigm. Its successes and failures should be particularly revealing about the nature of these new land management responsibilities, where and why current reforms have failed and, moreover, how future policies and programs might be structured to facilitate more decisive decision-making. Third, the national forest system lands are comprised of densely forested mountains and valleys and snow-capped mountain peaks. These lands are exceptionally scenic and are treasured for their breathtaking vistas, backcountry recreation experiences, and opportunities for solitude. And, because they are relatively undeveloped, national forests provide important wildlife habitat. At the same time, these lands harbor a significant share of the nation's timber, minerals, and fuels resources. As a result, they pose tremendously difficult resource allocation decisions for USFS officials and provide particularly vivid conflict for analysis in this context. Finally, the Forest Service has been chosen for analysis because the impasse created by these disputes in that agency, as will be seen, is severe. The agency is spending more than $30 million each year comprehensively planning and managing their resources in order to resolve these disputes, but the disputes and, moreover, their consequences, persist. (U.S. General Accounting Office, 1978)

OIL AND GAS LEASING AND PERMITTING

The process under which decisions are made governing where and how oil and gas exploration and development may occur on public lands was established in 1920 as a result of the efforts of progressive conservationists. Like other public land policies at that time, the Mineral Leasing Act of 1920 (30 U.S.C. 181) was a response to undesirable mineral development practices and was intended to improve conditions by regaining federal control. The oil and gas leasing and permitting programs developed by this act were meant to overcome the then pervasive fraud, wasteful production practices, and monopoly control over these resources. The objective of this new policy was to make energy production on public lands more orderly and efficient by controlling the *rate* of production. The appropriate decision-makers were deemed to be the professionals then staffing land management agencies (Ise, 1926; Noble, 1982).

When the Mineral Leasing Act was passed in 1920, Congress established a process by which the oil and gas resources contained beneath the public lands were to be leased and extracted. For more than 50 years, this process worked effectively with very little problem. As with other public land decisions, however, the political environment within which oil and gas exploration and development decision-making occurs has changed since the 1920s. Now these decisions must be considered in concert with many other, often conflicting, natural resource objectives. Decisions that previously would have taken a month or two under the leasing process are now taking years to make. Decisions that previously would have involved only the Bureau of Land Management or the U.S. Forest Service and a single lease applicant, now involve the Secretary of the Interior, Congress, the courts, numerous interest groups, individual citizens, and even the President. Regardless, oil and gas decision-making continues to be guided by the 1920 Act in the context of the land management paradigm devised during the progressive era of conservation. And, as with other issues, decisions regarding oil and gas are frequently unable to withstand the attacks leveled against them.

There was a time when the national forests were used almost exclusively for timber, watershed protection, wildlife management, agriculture, grazing, and recreation. In fact, until Michael Frome's updated analysis of the agency appeared in 1984, the extensive literature on the U.S. Forest Service seldom mentioned oil and gas or other minerals (Culhane, 1981; Dana and Fairfax, 1980; Frome, 1974b; Robinson, 1975; Steen, 1976). Suddenly, in the late 1970s, oil and gas leasing became classified as "one of the most sensitive concerns that the Forest Service deals with," according to John Crowell, former Assistant Secretary of Agriculture (Crowell, 1981). It fuels a controversy that the *New York Times* describes as "complicated, even nasty . . . no matter which side wins, the outcome is likely to be irrational" (Editorial, August 25, 1981). It is a controversy that leads a U.S. Forest Service Supervisor to comment that "a lot of people are watching to see just how bloody we get as a result of this thing. But, whichever way we go, I'm afraid we'll end up in court" (*New York Times*, August 30, 1981). It is a controversy that leads the National Audubon Society's vice-president to proclaim that "we prefer to work together in harmony . . . but, if war is forced upon us we will fight back!" (Evans, 1981). Disputes about where and how oil and gas exploration and development should occur have led to appeals and lawsuits that have decision-making on over a thousand leases at an impasse (*Congressional Record*, May 20, 1980, S5650).

TIMBER HARVESTING AND MANAGEMENT

Timber harvesting and management is one of the main responsibilities that originally gave rise to the U.S. Forest Service. As is discussed in Chapter 2, the disastrous "cut and run" practices of private lumber interests during the mid- to late-1800's destroyed valuable timber land and created severe fire and flooding problems. As various national forests were reserved from the then accepted land disposal programs, the USFS was established to manage these lands in order to

> improve and protect the forest . . . for the purpose of securing favorable water flows, and to furnish a continuous supply of timber for the use and necessities of citizens of the United States. (16 U.S.C. 475)

For over 60 years, agency foresters fulfilled this mandate with little controversy. Since the 1960s, however, this situation has become reversed. Controversies have arisen over where and, moreover, how timber harvesting should occur, and, additionally, over what size timber sales are appropriate. USFS silvicultural practices have been called into question. As a result, the USFS is suddenly finding itself in the midst of disputes that frequently undermine agency officials' ability to make viable decisions. Because of various administrative appeals and lawsuits, timber-sale decisions in several national forests have been at an impasse since the mid-1970s. As will be seen in the following chapters, USFS officials have undertaken several studies and have proposed several recommendations designed, unsuccessfully, to end these conflicts. After more than a half century of relative quiet in the national forest system, timber management and harvesting practices demand the attention of federal policymakers.

This impasse in timber management decision-making, created by appeals and lawsuits, has those private timber companies that are dependent upon public timber stands unable to plan their production activity. George Weyerhaeueser, president of Weyerhaeuser Company, believes that the Forest Service has lost the ability to make decisions that can be stood by. He argued before the Society of American Foresters' 1979 Convention that the agency is "caught in a cycle of legislate, plan, hold hearings, legislate again, litigate, legislate again to adapt to or overturn court decisions, and then begin the planning and hearing cycle once more" (*Journal of Forestry*, 1979, 77(1), p. 8). The U.S. General Accounting Office leveled a similar charge against the Forest Service. With the agency spending upwards of $30 million per year on its planning functions, this government watchdog agency queries why the Forest Service cannot anticipate the extent to which administrative appeals and lawsuits continually undermine its ability to reach forecasted timber production targets (U.S. General Accounting Office, 1978). The situation is little different in 1987.

NATIONAL FOREST PLANNING

In 1973, the late Senator Hubert H. Humphrey introduced legislation that eventually became the Forest and Rangeland Renewable Resources Planning Act of 1974 (16 U.S.C. 1601). At that time he described to the Senate the situation in the national forests with the comment that "to put it bluntly, we have a mess on our hands" (The Wilderness Society, Sierra Club, National Audubon Society, Natural Resources Defense Council, Inc., National Wildlife Federation, 1983, p. 9). In 1976, Senator Humphrey again went before the Senate in search of a solution to the continuing impasse over timber harvesting in the national forests. He spoke in favor of amendments to the 2-year-old resources planning act, amendments that he claimed were needed "to get the practice of forestry out of the courts and back to the forests" (Hall and Wasserstrom, 1978, p. 523). As a result, the National Forest Management Act (NFMA) of 1976 was enacted (16 U.S.C. 1600(note)). In late 1983 and early 1984, as the first handful of forest plans mandated by this forest management act were being released, they were greeted by the same lawsuits and administrative appeals that they were meant to replace.

At the outset of this NFMA planning process, Forest Service officials in the Washington, D.C. office expected a total of 200–300 administrative appeals of final plans on all 155 national forests. By April 1987, 600 appeals had been filed on the first 75 forest plans completed, 39 alone on Montana's Flathead National Forest. And, according to the agency's public involvement coordinator, these first 75 plans have been "the easy ones." The agency now estimates that there will be between 1200–1300 appeals before the planning process is complete (Yonts-Shephard, personal communication, April, 1987).

SIMPLE EXPLANATIONS BUT NO SIMPLE SOLUTIONS

Despite repeated and varied attempts by the USFS to quell the controversies generated by their forest planning efforts, timber sale proposals, and oil and gas exploration decisions, decision-making remains at an impasse. There are many explanations offered for why these disputes exist and persist. The forest products and oil and gas industries place the blame on environmentalists as well as on government regulation. They call environmental groups "extremists" intent on "locking up" public lands for the select few at the expense of the general and needy public. (Overton, 1981) Additionally, government regulation is chastized for placing obstacles in the way of needed energy development and for withdrawing lands critical to a healthy timber and home-building economy.

The oil and gas industry feels "double-crossed" by a mineral leasing process that, on the one hand, encourages the development of domestic energy resources while, with the other hand, places "one hurdle after another" before them and thus limits the exploration and development activities that can occur. The industry believes that "federal leasing means no leasing" (Edwards, 1981). Industry representatives cite case after case where "permitting and leasing delays . . . have held up drilling as much as 5 years." They are frustrated with "environmental rules [that] stifled development in many areas, wilderness studies [that] discouraged activity and threatened to lock up millions of acres, and leasing delays [that] discouraged exploration in some of the best onshore areas" (*The Oil and Gas Journal*, November 10, 1980, p. 140). Similarly, timber industry spokespersons question the extent to which "environmental impact statements, roadless studies, primitive area studies and possible wilderness classifications have slowed the Forest Service in their timber sale programs" (*Living Wilderness*, April/June 1976, p. 21). Their representatives criticize a "regulatory swamp" that has bogged down timber production from national forests in the last decade. A Montana lumber company president claims that "destruction of the forests will come from locking them up, not from cutting them" (*Living Wilderness*, April/June 1976, p. 22).

Environmental groups argue to the contrary. They criticize timber harvesting and oil and gas leasing decisions that threaten established land-uses and noncommercial resource values. Industry, they argue, is forcing these confrontations by its efforts to develop pristine wilderness and treasured national forest areas. They believe that industry has contrived its energy and economic "crises" to rape and ruin America's wild lands. William Turnage, executive director of The Wilderness Society, argues that industry's proposals would impose "an industry lock up" that would "sacrifice the country's last unspoiled wildlands for the profit of a small number of corporations" (*Living Wilderness*, Summer 1981, p. 44). If left untethered, they argue, industry will develop the last maining wild places. Environmentalists perceive it to be their responsibility to help curb this "blind progress."

Environmental groups echo industry displeasure with the administrative decision-making process. They criticize how Forest Service officials fail to respond to the issues that environmental groups raise: "Like all old bureaus, the once highly respected Forest Service has grown inflexible. It has been unable to comprehend or respond to the change in public attitude toward the environment" (*Living Wilderness*, Summer 1981, p. 44). Environmental organizations question not only the substance of Forest Service decisions, but also the process by which these

decisions are being made: "Never in recent history have conservationists felt so frozen out of decision-making on public lands" (Reilly,
1981).

Given their dissatisfaction with the formal administrative decision-
making process, industry and environmental groups are both encouraging Congress and the Secretaries of Agriculture and Interior to take
action to end the impasse. Unfortunately, the nature of the problem
evades such simple solutions. During the Carter Administration, it was
thought that deferring controversial oil and gas leasing decisions would
stifle that conflict. Carter's Secretary of the Interior Cecil Andrus found
himself in court as a result of this inaction (*Rocky Mountain Oil and Gas
Association* v. *Andrus*, 550 F. Supp. 1338; *Mountain States Legal Foundation*
v. *Andrus*, 499 F. Supp. 383). Under the Reagan Administration, industry believes that it has the upper hand. Robert Nanz, vice-president of
Shell Oil Company, calls for the opening and development of public
lands. He asserts that "Congress and the administration have the power
to get all this done quickly" (*The Oil and Gas Journal*, September 29, 1980,
p. 58). However, Reagan's first Secretary of the Interior James Watt's
management objective to "open wilderness areas" likewise landed him
in court. Congress has had no better luck. Its rapid action to prohibit
leasing in Montana's Bob Marshall Wilderness Area led it to court (529 F.
Supp. 982). Regardless of who tries to put the issue to rest or who
benefits by the decision made, the outcome is predictable. Like the
formal administrative process, these avenues for influencing decision-
making seem unable to resolve the pervasive conflict and thereby make
viable decisions.

Congressional efforts to resolve public land disputes have not been
limited to oil and gas issues. As discussed earlier and as will be illustrated in later chapters, Congress' attempts to take "the practice of forestry out of the courts and back into the forests" has not been entirely
successful (Hall and Wasserstrom, 1978, p. 523). Similarly, Congressional efforts to resolve the long raging conflict between conservationists
and preservationists over wilderness designations, have been ongoing
since the 1950s, and will likely continue into the twenty-first century. In
1977, Congressman Morris Udall, chairman of the House Interior and
Insular Affairs Committee, predicted that it was going to take between 2
to 4 years to finally resolve the wilderness allocation disputes. He lamented this delay in decision-making but justified it because each wilderness area was going to be so "intensely fought over." Few wilderness decisions were made then during the 96th Congress because, as
one senior Congressional aide explained, the decisions presented a "pot
of boiling oil that nobody is eager to jump into right now." James

Crowell, former Assistant Secretary of Agriculture during the Reagan Administration and having jurisdiction over the USFS, severely criticized past efforts to make wilderness designation decisions. He was concerned that these decisions be made quickly in order to avoid continued and costly uncertainty regarding the natural resources in these areas:

> For the last ten years, arguments over roadless areas have diverted attention and resources from other important long-term management issues. Areas have been studied and restudied. . . .
>
> If we do not settle the roadless area question once and for all, there is a risk that lawsuits can be brought whenever the Forest Service makes a decision to manage a roadless area for a nonwilderness purpose. This will perpetuate the cost, delay, and uncertainty of the last ten years—and perhaps still not guarantee the sanctity of the areas which truly deserve wilderness designation. (*American Forests*, March 1983, pp. 60–61)

After several years trying to reach agreement on Wyoming's wilderness allocation, Senator Richard Cheney (R-Wyo) vents his frustration with the comment that "my main desire in life is to get one [a Wyoming Wilderness Bill] done" (*High Country News*, December 12, 1983, p. 1).

As an example of the extent to which conflict pervades wilderness designation decision-making, consider the Cougar Lakes case. After evaluating the wilderness potential of the Cougar Lakes region of Central Washington, the USFS proposed that 23,000 acres be designated as wilderness. The National Park Service, in contrast, evaluated the wilderness potential of the same area and proposed that a 138,000-acre wilderness be established. Following these agency analyses, the Carter administration sent Congress its proposal that a 129,000-acre wilderness be established. Encountering such divergent proposals, Congressman Mike McCormack (R-WA) proposed his own compromise solution: a 38,600-acre wilderness, and a 53,600-acre "limited recreation area" in which some timber harvesting and road building would be allowed. However, two other Washington Congresspersons argued that "a more even balance between Wilderness and nonwilderness" was needed in Washington State. They introduced another bill proposing a 270,000-acre Wilderness (Robinson, 1979, pp. 24ff).

ADMINISTRATIVE DECISION-MAKING AT AN IMPASSE

That national forest management decisions generate conflict has become an undeniable fact. And, although this conflict is acknowledged and anticipated, the administrative decision-making process is unable to provide direction on how to manage and resolve it. Consequently,

federal land managers are unsure about how to fulfill their responsibilities. Decision-making has become confused, at best. In Wyoming's
Shoshone National Forest, newly appointed Forest Supervisor Steve
Mealey found himself facing a lawsuit by the Foundation for North
American Wild Sheep over a decision made 2 years earlier by his predecessor. Mealey agreed with the plaintiffs contentions and decided to
reverse the earlier decision concerning seismic exploration until a new
environmental assessment could be completed. Upon making this decision, however, Mealey found himself confronting an administrative appeal to overturn it filed by the International Association of Geophysical
Contractors. Whatever decision Mealey makes only moves him into a
different arena for administrative or judicial review (*High Country News*,
November 28, 1983, p. 4).

Oil and gas exploration and development, an area that is only recently posing management difficulties for the USFS, provides particularly vivid examples of the administrative difficulty of current land management responsibilities. For example, contrast how leasing decisions
were made in three different U.S. Forest Service regions during 1981.

1. In New Mexico, lease applications were filed for the Capitan
Wilderness area. The Regional Forester of the Southwestern Region
viewed the leasing decision to be an insignificant one, involving no
environmental impact and therefore requiring no public notification or
environmental assessment. The leases were issued. When representa-
ives of the Sierra Club discovered that the leases had been issued, they
were outraged. The environmental organization immediately went to
Congress to protest the decision. Their efforts resulted in a bill being
introduced in Congress to withdraw *all* wilderness lands from oil and
gas leasing. The Sierra Club additionally filed suit again the U.S. Forest
Service, Bureau of Land Management, and Secretary of the Interior to
revoke these leases (*New England Sierran*, May 1982, p. 5).

2. In Montana, lease applications and seismic testing permit applications were filed for the Bob Marshall Wilderness Area. The Regional
Forester of the Northern Region viewed the decision to be so significant
that he denied the leases and permits (*High Country News*, May 29, 1981,
p. 14). When the Mountain States Legal Foundation, an industry interest group, sued the U.S. Forest Service to overturn this decision,
national environmental organizations began pressuring the U.S. House
Committee on Interior and Insular Affairs to take action. The Committee
held a brief hearing and concluded that an emergency existed in the Bob
Marshall Wilderness Area because of the pending lease applications
there. They invoked an emergency provision of the Federal Land Policy
and Management Act of 1976 to withdraw the area from all oil and gas

exploration and leasing. The Mountain States Legal Foundation and Pacific Legal Foundation brought suit against the Congressional Committee, contending that the withdrawal was unconstitutional because the FLPMA section used by the committee was itself unconstitutional, violating the separation of powers clause of the U.S. Constitution (*High Country News*, May 29, 1981, p. 2; June 12, 1981, p. 2).

3. In Wyoming, the Rocky Mountain Region Regional Forester took what he viewed to be a more balanced approach to decision-making. In response to lease applications in the Washakie Wilderness Area, an environmental impact statement was developed and the conclusion reached that 87% of the area should *not* be leased while 13% could be leased. This balanced decision, rather than pleasing all sides, has both industry and environmental groups dissatisfied and preparing administrative appeals and lawsuits to overturn the decision once finalized (U.S. Department of Agriculture, Forest Service, 1981).

In each case, no matter how the decision was perceived (as extremely significant or inconsequential), no matter who would benefit from the decision (environmentalists or the oil and gas industry) and no matter what type of analysis was completed (no assessment to a full environmental-impact statement), the outcome was the same: conflict resulted and the debate was carried on to the courts, Congress, or the administrative appeals process for further review. None of the decisions was able to accommodate the concerns of each group to their satisfaction. Each decision prompted opposition rather than acceptance and support. Each case involved considerable expense to all parties involved.

THE APPROACH OF THIS BOOK

Before new policies or programs are generated or new charges leveled, the problem currently encountered in making public land management decisions must be explored and understood. The next seven chapters analyze the policy problem posed by public land management in the 1980s. They place three issue areas—timber harvesting and management, oil and gas leasing and permitting, and national forest planning—under the microscope to determine why existing conflicts are not being resolved by current administrative processes, and, moreover, what the consequences are of this failure. In so doing, the nature of public land management today is better understood and the type of policy process best suited to this type of policy problem identified. This analysis concludes by suggesting specific administrative processes that might be employed to better accommodate all affected interests in public land management decision-making, and, thereby, to reduce the now inevitable opposition to land agency decisions.

Chapter 2 reviews the history behind the reservation and management of the national forests. It describes how and why various administrative processes have transformed today's approaches to problem-solving. It explains the development and maturing of the forestry profession and its effect on national forest management. The undercurrent throughout the chapter is how the public land management paradigm was developed and adopted, and what its impact has been on how issues are analyzed and decisions are made governing the national forests.

Chapter 3 presents the administrative procedures followed by Forest Service officials at all levels in the agency in decision-making. It describes what is at stake in these decisions as well as what mitigation measures agency officials can use to minimize adverse impacts. It pinpoints where professional discretion is exercised, where public input is acquired, and how various public land-user groups are involved.

Chapter 4 analyzes the land management paradigm in practice in the context of three issue areas chosen for analysis. It explores the political dimensions of what are undeniably very political processes. It identifies three critical pathologies that contribute to the paradigm's shortcomings: the process is not sufficiently informative or convincing, it is divisive, and it is not decisive.

Chapter 5 determines where and why the paradigm now fails, when for so long it was very successful. It discusses the changing public sentiment regarding the appropriate use and management of public lands in the context of a changing social and political climate. It describes several natural resource statutes that have expanded the objectives to be satisfied in land management, and why the paradigm is not able to accommodate them.

Chapter 6 asks why this mismatch between policy problem and process persists. It reviews the findings and arguments of other students of public land management who have identified this problem and who have proposed solutions. Moreover, it highlights USFS efforts to remedy the situation. It pinpoints where these proposed and attempted reforms have succeeded, and where they have failed.

Chapter 7 describes the history and intent behind the National Forest Management Act of 1976, an act that many hoped would facilitate decisive decision-making by resolving the disputes highlighted in this book. Building upon the lessons of Chapters 4 through 6, it proposes five key elements that should be integral to any process designed to foster effective, viable decision-making: building *trust* among affected parties and decision-makers; building *understanding;* incorporating conflicting *values;* providing opportunities for *joint fact-finding;* and, encouraging *cooperation* and *collaboration* among affected parties.

Chapter 8 takes the lessons of earlier chapters and describes the elements of a process that might address these failings, a process designed specifically to manage the inevitable conflict and resolve specific disputes whenever possible. It illustrates how the key elements of this process have been successfully applied in various national forest management situations, and why further experimentation is warranted.

Finally, Chapter 9 concludes the book by presenting an experimental eight-step process, currently being followed in some federal agencies, that the Forest Service and Secretary of Agriculture might follow to begin tackling the public-land management problem of the 1980s.

CHAPTER 2

The Historical Context of Contemporary Forest Management

The federal government controls one-third of the nation's land—740 million acres. These "public lands" contain tremendous and varied resources. Some of these resources are found on the surface: timber, grazing and agricultural lands, wildlife habitat, recreational and scenic amenities, and wilderness. Others lie beneath the surface: coal, potash, phosphate, sulfur, oil shale, helium, copper, and oil and gas. The resources comprising the public lands are managed by several different federal agencies. The Department of the Interior has jurisdiction over public lands through its Bureau of Land Management (398 million acres), Fish and Wildlife Service (43 million acres) and National Park Service (68 million acres). The Department of Agriculture, through the U.S. Forest Service, controls 187.5 million acres. The remaining acreage is split amongst many different agencies including the Departments of Defense, Transportation, and Energy.

The history of the public lands—their acquisition, disposal, and the gradual development of laws governing their management—is hardly a dull story. It is a story of the taming of the "Wild West" and the forces of "good" against "evil." It is a story of a relentless struggle for power and wealth. The history leading to today's public-land management process is one rife with political battles in Congress, the courts, and federal agencies. It is a history of land management agencies fighting to maintain their administrative discretion and therefore their ability to make professional decisions.

19

THE HISTORY OF PUBLIC LAND MANAGEMENT

Public land policy has evolved through three major phases. The first phase (1781–1867) covered the massive acquisition of land by the United States from the original thirteen states, through purchase from European monarchies (i.e., the Louisiana Purchase and the Alaska Purchase) and through transfers from Mexico. Through these acquisitions, four-fifths of the nation's land were at one time or another under the jurisdiction of the federal government. The second phase (1872–1934) saw the massive disposition of these same lands by the federal government to promote development of the new territories and generate revenues for the General Treasury. Indeed, the Treasury Department was the nation's first public land manager. The third and current phase (1891–present) is characterized by a shift from disposal of public lands to federal government retention and management (Clawson, 1971, Chapter 1 and pp. 26–28; Culhane, 1981, pp. 41–43).

For the purpose of this book, the important time period in federal land management is the late nineteenth and the early twentieth centuries. During this time the federal government shifted from a policy of disposal in hopes of encouraging development to a policy of retention and conservation. It was a time when the government began to acknowledge the commercial value of these lands; a time when predominant public sentiment, influenced by scientific arguments, embraced the view that federal ownership and management could more directly serve national needs than could disposition.

PUBLIC LANDS IN THE LATE-NINETEENTH CENTURY

As the nineteenth century came to a close, it was readily apparent that the federal government's public domain disposal programs were contributing to wasteful development and lawlessness in the West. Rather than "taming" the wild West, these programs were encouraging its destruction. The government's policies had not been well-conceived and they required reassessment and revision.

Many "settlers" were using the homestead laws to acquire public lands for nonagricultural purposes (Clawson, 1971, pp. 15–16; Hays, 1959; Peffer, 1951). With disposal programs run from Washington D.C., 2,000 or more miles away from the West, there was little field surveillance. Most federal employees in the General Land Office (GLO), in fact, had never even seen the lands they were disposing. Often, individuals acquired land intending only to speculate, not to develop, its resources. In 1889 alone, 55 General Land Office agents spent the equivalent of 30 work-years investigating 3,307 cases of land fraud. Timber trespass accounted for 581 of these cases and $3–$6 million in lost revenues to the federal government (Steen, 1976, p. 25). Timber fraud was widespread. One GLO land agent estimated that 75% of the timber claims he reviewed were fraudulent. A less pessimistic agent put the figure at 50% (Steen, 1976, p. 25). Illegal entry by individuals whose intention was to cut timber or extract minerals was rampant.

Timber cutting practices by private lumber companies reflected the bountiful supply of forests and lack of management principles. Private companies paid considerably less than market prices for public timber. As a result, only first-rate timber was taken. The still marketable second-rate timber was left as waste as lumber companies moved on to the next forest. These "cut-and-run" practices created extreme fire danger and disease problems throughout the forests. In 1871, a single forest fire destroyed the town of Pestigo, Wisconsin, killing 1500 people and burning 1 million acres of forest (Culhane, 1981, p. 4; Hays, 1959, p. 27). Additionally, the failure of private companies to reseed and restore cut areas increased erosion and led to flooding in some areas (Culhane, 1981, p. 47). These visible consequences of mismanagement eventually helped professional foresters gain support for forest retention and management and helped to institutionalize two goals of forestry management: watershed protection and fire prevention.

The western range hardly fared better than the forests. With free grazing lands everywhere, cattle and sheep herds rapidly multiplied. The classic "tragedy of the commons" led to overgrazing in major areas with substandard forage replacing the original vegetation (Clawson, 1971, pp. 71–72; Culhane, 1981, p. 81). Some stockmen took the liberty to fence off portions of the range for their own use, but rival stockmen would inevitably break down those barriers (Hays, 1959, pp. 49–53).

The homestead programs were failing because they had been developed without an understanding of western agricultural conditions. Instead, the programs were based upon eastern farming needs. The West, being very arid, required much larger acreages than in the East to make workable farms. Unlike the East, irrigation programs were necessary in the West (Culhane, 1981, pp. 80–81). The more promising agricultural areas were also the better grazing lands. When settlers did move into an area to farm under the Homestead Laws, violence erupted between settlers and the stockmen already grazing there (Hays, 1959, p. 53).

Finally, it was becoming obvious that the best western forests, range, and agricultural lands were rapidly being acquired by private interests, and that the federal government would soon be left, literally, with "the lands nobody wanted" (Clawson, 1971, pp. 18, 69). Indeed, when the dust finally settled, what remained were the more marginal mountainous forests and the uncultivatable, substandard range, and desert. To a large extent, this condition still persists; after decades of management, 83% of BLM range lands are still considered substandard for grazing purposes (Culhane, 1981, p. 95).

Both the federal government and private land users acknowledged the deteriorating condition of the public domain. Cries for reform came from many fronts but most forcefully from preservationists and conservationists. Preservationists, who viewed *any* development of some public lands as undesirable, successfully fought to have Yellowstone National Park set aside as the first national park in 1872. One historian wrote of their efforts: "The bill to establish Yellowstone succeeded after one of the most formidable, public-interest lobbying campaigns in history" (Frome, 1974a, p. 19). Preservationists continued their efforts to preserve scenic forest lands and, in 1891, were able to obtain an amendment to the General Land Law Revision Act that gave the President authority to create forest reserves by proclamation (Hays, 1959, p. 36). The act contained no provisions for the management of these forests once reserved, however.

Forest Reserves and National Parks could not remedy public domain problems on a broader scale. The federal government continued to be preoccupied with the development, and not the preservation, of the West. The conservationist's alternative of scientific land management seemed to provide the solution to the land problems then plaguing the West.

THE PROGRESSIVE ERA OF CONSERVATION

The history of the conservation movement has been well-documented by historian Samuel P. Hays (1959). Hays rejects the traditionally accepted belief that widespread protest and unified support gave birth to the conservation movement. Rather, he argues that "it is from the vantage point of applied science, rather than of democratic protest, that one must understand the historic role of the conservation movement" (p. 2):

> Conservation, above all, was a scientific movement, and its role in history arises from the implications of science and technology in modern society. Conservation leaders sprang from such fields as hydrology, forestry, agrostology, geology, and anthropology. Vigorously active in professional circles in the national capital, these leaders brought the ideals and practices of their crafts into federal resource policy. Loyalty to these professional ideals, not close association with the grass-roots public, set the tone of the Theodore Roosevelt conservation movement. Its essence was rational planning to pro-

mote efficient development and use of all natural resources (Hays, 1959, p. 2).

The professional orientation of the conservationists was not universally accepted nor immediately adopted by Congress. To implement the conservation ideals in their purest form would entail wresting power from the commercial interests already entrenched in the West and, in turn, placing it in the hands of professionals dedicated to conservation in land management. It required tremendous political support to counter the already well-supported western interests. Congress needed reason to change. At the time, the land-disposal programs, combined with tariffs and excise taxes, generated most federal revenues (Steen, 1976, p. 4).

Success came slowly, but was facilitated by a parallel social reform movement that advocated management principles for municipal and industrial management. The nation was entering the progressive era of political reform; values towards the appropriate role of government, science, and industry in society were being transformed. Society was entering a technological age wherein scientific methods of management and decision-making were perceived to be a godsend that would improve living conditions, create efficiency, and overcome political corruption in urban centers, as well as in the West (Beard & Beard, 1960, pp. 368–383). Society was also leaving the frontier period during which rapid growth and development were both desired and rewarded (Wiebe, 1967, p. 167). Now it turned to managing the cities and the public domain and to gaining control over those who had exploited them in the past; The Sherman Anti-Trust Act was passed in 1890. Groups such as the American Civic Association, General Federation of Women's Clubs, and Daughters of the American Revolution were actively pursuing similar objectives in the nation's cities, trying to wrest power from politically corrupt influences and install administrative structures based on theories of efficient municipal management, service provision, and rational planning. As Hays (1959) notes:

> The conservation movement was closely connected with other organizations which attempted to promote efficiency. Leaders of the Roosevelt administration, for example, maintained close contact with the four major engineering societies, the American Society of Civil Engineers, the American Society of Mechanical Engineers, the American Institute of Electrical Engineers, and the American Institute of Mining Engineers. The societies spearheaded the drive for efficiency. (p. 123)

Similarly, the American city-planning profession was forming. These groups gave their support to the natural resource conservation movement.

Progress in institutionalizing the conservation ideal was fostered by the respected and influential standing of its advocates. As scientists and economists, they were typically well-bred, well-educated and traveled in the same circles as the Congressmen they sought to influence (Steen, 1976, Chapters 1–4). The country was in a period during which knowledge and science were valued and respected. This attitude was reflected in Congressional decision-making. As Beard and Beard (1960) describe it:

> By that time it had also become a practice for Congress and state legislatures to appoint special committees authorized to employ experts and carry on extensive investigations, before undertaking the business of lawmaking relative to such intricate matters as the regulation of railways, conservation of natural resources, and the provision of social security. (p. 338)

And, obviously, those appointed to these commissions were those who had initially raised the issues and potential solutions: the professional, applied-science community.

Although these early conservationists set the scene for change, their scientific arguments could not alone effect that change. What was needed was political savvy. Success would not be achieved until the strong western opposition could be quieted or overcome—that implied compromise. Moreover, Congress needed to be convinced that land management served its already-articulated objectives in western development and revenue generation. The conservation principles needed to be presented in a manner that conformed to already-defined objectives. New objectives could be proposed only if they did not conflict with those already established.

In the 1890s, the conservation ideal in public land management began to take hold through the dynamic personality and relentless efforts of its natural leader and spokesperson, Gifford Pinchot. Pinchot was a forester by training. Although educated at Yale, he gained his practical experience in Germany and France. He brought back to the United States the forestry skills perfected in Europe and, after a brief period managing a private forest in North Carolina, was convinced that current silvicultural, watershed management, and fire-control methods could be applied to the public timberlands to prevent the widespread disease, flood, and fire problems there (Hays, 1959, pp. 28–29). Pinchot's objective, and one that has been the driving force behind professional land management efforts ever since, was to manage the public lands to achieve, as he put it, "the greatest good of the greatest number in the long run" (Pinchot, 1947, pp. 319–326). Conservation of resources, he argued, was the "basic material problem of mankind" (Steen, 1976, p. 255). Conservation was the efficient, sustained use of public land resources to serve the nation for all time. Bernard Fernow, Pinchot's predecessor as Chief of the Division of Forestry, described it as the efficient use of the interest from public land resources, while never having to dip into the capital (Steen, 1976, 38). Pinchot, Fernow, and their colleagues were so convinced of the appropriateness of conservation principles that they referred to it as "a question of right and wrong" (Wiebe, 1967, p. 153).

In 1897, Congress passed the Forest Management Act. The Forest Management Act was the first legislative mandate for land *management* rather than disposition or unmanaged reservation. It gave the Secretary of the Interior power to "make such rules and regulations and establish such services as will insure the objects of [the] reservations, namely, to regulate their occupancy and use and to preserve the forests therein from destruction" (Wiebe, 1967, p. 44). Future reserves could be established "to improve and protect the forest within the reservation, or for the purpose of securing favorable conditions of water flows, and to furnish a continuous supply of timber for the use and necessities of citizens of the United States" (Steen, 1976, p. 36). The act was a tremendous victory for Pinchot and his colleagues, who argued that "legally there is no obstacle to the introduction of the most practical and approved ways of handling forest lands" (Hays, 1959, p. 44).

The wheels of government turn slowly, though. Passage of the Forest Management Act in 1897 and Pinchot becoming Chief of the General Land Office's (GLO) Division of Forestry in 1898 did not pave the way for the widespread and unhindered application of professional forestry to the public lands. It took time for Pinchot to overcome the obstacles posed by the already existing GLO and its staff of law clerks "trained in the legal details of land disposal but thoroughly unfamiliar with forestry or the West" (Hays, p. 37):

> Trained as lawyers, they had no large views of the possibilities of forest management, but adhered strictly to narrow interpretations of law and emphasized formal procedures rather than results. The custom of political appointments to the General Land Office hampered the selection of technicians. Politicians considered the position of forest supervisor as a patronage plum, for example, and bitterly criticized the General Land Office when it selected trained men for the post. (Hays, 1959, p. 37)

Pinchot persevered. He mobilized professional foresters and forestry associations behind him, including those in the private sector. Pinchot immediately instituted fire control programs and the selective cutting of timber in the forest reserves. His successes encouraged large private timber corporations that had much to gain by a managed and sustained timber industry and he thereby gained their backing. For the first time, the Division of Forestry was staffed by professional foresters (Hays, 1959, p. 46).

Pinchot found a natural ally in Theodore Roosevelt whose efforts had helped preserve Yellowstone National Park. Roosevelt, as one historian wrote, joined those political causes in which he found "personal relevance" (Wiebe, 1967, p. 190). Because Roosevelt was a conservationist, the conservation ideal made historic strides during his administration. To no insignificant degree, this progress was due to Pinchot's and Roosevelt's close friendship. (This, of course, was neither the first nor the last occasion when a social problem was addressed because of the personal interest or need of a powerful political actor [Kelman, 1980]). Pinchot's diaries recount the many occasions when he and Roosevelt discussed conservation and strategized its implementation (Pinchot, 1947). Pinchot's favorite account was when he, Roosevelt, and the French Ambassador eluded secret service men and "stripped to the buff" to go swimming in Washington's Rock Creek Park (Steen, 1976, p. 70).

Pinchot and Roosevelt were fortunate to be pressing these concerns at a time when social reform was rampant and Congress was receptive to the widespread "gospel of efficiency." Roosevelt saw much to be gained by the new theories of public administration that were developing at the time. He believed that the executive office was the "direct representative of the people at large," and therefore it was the executive's responsibility "to guide the public and Congress toward that resource policy best for the entire country" (Hays, 1959, p. 134). Roosevelt sought a more efficient federal administration. However, he realized that "to make it so is a task of complex detail and essentially executive in its nature; probably no legislative body, no matter how wise and able, could undertake it with reasonable prospect of success" (Hays, 1959, p. 134). He perceived administrative decision-making divorced from the political dealings of Congress to be the only way to achieve efficient and rational land planning. Historian Robert Wiebe (1967) has attributed Roosevelt with inverting the traditional relationship between the executive and Congress:

> Now the subdivisions of the executive had assumed the task of studying and resolving the big problems. The President was expected to give priorities, then focus Congressional attention on an issue at a time; in other words, provide and direct a rather precise legislative program. The President initiated, and Congress, if it wished, could veto. (p. 193)

Pinchot was fortunate to have Roosevelt's dedication to the conservation movement. Pinchot could suggest policies or legislation and Roosevelt would see to it that Pinchot's suggestions became a reality. During the Roosevelt administration, the forest reserves increased by 148 million acres to total 194.5 million acres (Robinson, 1975, p. 9). In fact, Roosevelt's enthusiasm in establishing forest reserves caused Congress to pass a law requiring Congressional authorization for all future reservations. Roosevelt had no choice but to sign this measure into law, because it was attached to a much needed appropriations bill. However, in the few hours before he was forced to sign it, Roosevelt established 16 new national forests (Steen, 1976, p. 100).

Although it certainly made significant strides, the Pinchot–Roosevelt alliance alone could not have imposed the dramatic change then occurring with respect to the public lands. Much of their initial success came from a willingness to accommodate western interests and thereby avoid an undoubtedly insurmountable obstacle to forest management. Some of the proposed regulation, by that time, was actually desired by western range and grazing interests. Wilson (1973) refers to this as the "self-interest" theory of regulation. By taking public timber off the market, except where needed, the forest reserves helped stabilize a severely fluctuating timber market (Steen, 1976, pp. 89–96). By systematically allocating grazing permits, the forest reserve range was better maintained, and stockmen were no longer competing with one another for the same foraging areas at the expense of the quality of the land (Steen 1976, pp. 86–89).

Other factors contributed to western acquiescence. "Primary consideration" in all Forest Service regulations was given to accommodating local interests (Steen, 1976, pp. 78–79). Settlers were still allowed to cut timber. The Forest Service *Use Book*, the 142-page precursor to today's more than 20 volume *Forest Service Manual*, stated at the outset the basic tenet of the Forest Service:

> Forest reserves are for the purpose of preserving a perpetual supply of timber for home industries, preventing destruction of the forest cover which regulates the flow of streams, and protecting local residents from unfair competition in the use of forest and range. They are patrolled and protected, at Government expense, for the benefit of the Community and home builder. (Steen, 1976, pp. 78–79)

Moreover, the first eight regulations detailed in the *Use Book* regarded "preferential treatment" for local users in obtaining free-use permits, and generally protecting their economic interests in use of the forest resources. Regulations regarding grazing in the national forests emphasized the dual intent to "contribute to the well-being of the livestock industry" and "protect the interests of the settler against outside competition" (Steen, 1976, p. 79).

With the national forest essentially reserved until such time that its timber would be needed to meet local and national demands, the Forest Service turned its attention to assisting private companies in the management of their lands. Forest Service staff were available, at private company expense, to develop long-term timber management plans for the sustained yield of timber from private forests. Other cooperative programs included tree plantings, fire prevention and fighting, and disease and pest control (Steen, 1976, Chapters 3, 5, 7).

One of Pinchot's major efforts during this time, and one immediately adopted by Roosevelt when he became President, was to obtain the transfer of all land-management responsibilities from the Department of the Interior to the Department of Agriculture. Pinchot argued that these tasks logically belonged to the Department of Agriculture because they were scientific in nature and based in the biological sciences. Moreover, he argued that "no one could cut through [the] entrenched inefficiencies" of the GLO that was "hopelessly involved in a maze of political appointments, legalistic routine, and personal favoritism" (Hays, 1959, p. 39). He believed that a new agency would be the only way to fully overcome the inefficiencies of the past and begin to professionally manage the public lands. He argued that all management responsibilities should be transferred because the resources themselves were interrelated and thus integrated administration would be the rational, efficient approach. Furthermore, it would overcome the problems caused by "interdepartmental conflicts which resulted when competing resource users played one agency against another for their own advantage." (Hays, 1959, p. 72). In 1905, Congress approved the transfer of the forest reserves, but not the Geological Survey, General Land Office, or Office of Indian Affairs to the Department of Agriculture. That act marked the birth of the U.S. Forest Service.

Progressive conservation was not pursued universally with the same zeal and enthusiasm exercised by Roosevelt and Pinchot. Roosevelt's successor, President William Taft, only mildly supported and sometimes opposed Pinchot's efforts. After the tremendous strides during the Roosevelt administration, progress in further institutionalizing the professional conservation ideal came slowly. It, nonetheless, did come. By that time conservation was, as one historian put it, "a common element of political rhetoric." President Taft reportedly lamented after his election that everyone was in favor of conservation, whatever it was (Steen, 1976, p. 96).

PRESERVING THE PARADIGM

The conservation movement was able to succeed at the turn of the century for three reasons. First, a problem existed. There was widespread fraud and abuse of the resources contained in the public domain, and, moreover, no means within established administrative institutions by which the federal government could control this abuse. Second, an organized and influential profession provided a rational solution to the problem. Conservationists had a United States President enthusiastically supporting their efforts, and they had scientific documentation for their claims. Third, the political and social climate at the time promoted efficiency and management in government, precisely what the forestry profession was advocating.

Although these factors explain why the ideal succeeded in 1900, they do not explain why it prevails in 1980. Several factors have contributed to the continuing success of the conservation paradigm. To a large extent, this success derives from the type of institution that Pinchot and his colleagues were able to set into place at the turn of the century. The U.S. Forest Service manages the national forests in a highly flexible and discretionary administrative environment. Pinchot argued for this flexibility in order to maintain "the high standard of fidelity, honesty and ability" needed to manage the forests, as well as to better enable Forest Service officials to function under such diverse natural conditions (Steen, 1976, p. 50).

Additionally, institutional strength came from strategic behavior on the part of Forest Service officials throughout this century. Although little legislation was actually enacted following the establishment of the national forests in 1905, the agency's powers continued to expand. As one Forest Service law officer later remarked:

> The vitalizing of this power through vigorous use was the chief means whereby the Forest Service achieved results in matters of grazing, water power, the prevention of land frauds, etc. Comparatively little conservation legislation was enacted during these years. Progress came not through getting new powers . . . but by using those we had. (Hays, 1959, p. 44)

Part of these results came through stretching legislative mandates and then receiving favorable rulings when taken to court. Forest Service historian Harold Steen (1976) observed:

> The Forest Service strategists understood well what many refuse to accept, that courts by their decisions do make law. By picking and choosing cases—and judges who were supporters of conservation—. . . [they] were able to build up an impressive corpus of common law to give substance to hard-won legislative battles. (p. 89)

Moreover, in contrast to today, judges at the time frequently preferred not to rule in matters that they deemed best left to administrative expertise. By not ruling, they validated the Forest Service's action and set precedent for future agency decisions (Brizee, 1975a,b).

The institution alone, however, does not account for the paradigm's strength today. The paradigm is not only a model of action premised in conservation ideals, it also represents the culture of the public-forestry profession. As a result, the profession itself has reinforced the paradigm through standard-setting for university-level forestry education, as well as through the esprit de corps of the agency. Throughout the twentieth century, there has been a revolving door between the Forest Service, faculty in forestry schools, and leaders in the professional forestry associations. Therefore, the paradigm has become entrenched in the profession's code of ethics and standards of behavior, and in Forest Service guidelines.

Additionally, the professional expertise model of decision-making persists because it was not really challenged until recently. Overall, the decades following establishment of the Forest Service were a rather sleepy time for the agency. It managed the forests as the early laws allowed and waited for the day when the nation would need timber from the public forests. However, the Forest Service was not without threats during this period. There were several efforts to move the agency from the Department of Agriculture (USDA) to the Department of the Interior (DOI). These proposals were always opposed by foresters and conservationists, fearing that it would bring about the demise of professional, nonpolitical forestry. These efforts failed when Congress and the President's attention was pulled to more pressing issues. BLM mismanagement of Alaska lands ended one transfer; World War II ended another (Steen, 1976, pp. 147–152, 237–245). To a large extent, these threats reinforced the paradigm as they increased tensions and name-calling between the foresters in USDA and land managers in DOI (Steen, 1976, pp. 237, 245). They helped sustain the esprit de corps that Pinchot believed was so critical to good forestry, by pulling together Forest Service staff as a team. Without this united spirit, there was no guarantee that loyalty to the agency's ideals would prevail in the field; the temptations to succumb to other pressures would be too great.

The greatest challenge to the paradigm came recently as Congress enacted several preservationist statutes. (These statutes are discussed in detail in Chapter 5.) Congress implicitly questioned the paradigm by supplementing conservation objectives with preservation and noncommercial objectives. However, while seemingly undermining the paradigm by legitimizing and empowering preservation ideals, Congress at the same time reinforced the paradigm by delegating extensive administrative powers to the scientific experts deemed best able to make these tough decisions.

CONSERVATION PRINCIPLES INVADE THE GRAZING LANDS

These dramatic changes in national forest management and the concommitant success of foresters in institutionalizing their professional land-management paradigm were not mirrored by land-management practices for the rest of the public domain. Land managers in the General Land Office adhered to a different philosophy regarding the proper use and disposal of public lands. Unlike the USFS, they believed that the land users themselves were the best able to dictate appropriate management measures.

The fate of the public domain apart from the national forests had been debated time and again in Congress and the courts, since before the turn of the twentieth century. It was not until the Taylor Grazing Act passed in 1934 that the long period of nonmanagement and mismanagement of these lands was brought to a close (43 U.S.C. 315). The act established grazing districts and fees for grazing in the open range. Placed in charge of the new Grazing Service in the Department of the Interior was Farrington Carpenter, a former rancher, Harvard Law School graduate, and one of the key witnesses during the Taylor Grazing Act hearings (Culhane, 1981, p. 84). Unlike Pinchot, Carpenter thought the issues to be addressed by the Grazing Service were best left to the stockmen themselves. He believed that federal officials who knew little about grazing and had never visited the western range should not make decisions that affected grazing. He established district advisory committees comprised of local ranchers to make recommendations to the Grazing Service. Carpenter regarded "practical range experience" to be the major qualification in hiring the district graziers who would act upon advisory committee recommendations. As a result, Grazing Service field staff were dominated by ranchers, not by professionally trained land managers as in the Forest Service. Carpenter's policy became known as "home rule on the range" (Culhane, 1981, pp. 85, 104). Because of Carpenter's loose control, the inferred power and independence of these advisory committees strengthened over the years. When a BLM range survey in the early 1950s determined that the allowable AUM (animal unit month—loosely defined as the amount of forage required to feed one mature cow for one month [Clawson, 1971, p. 37]) level in one district should be decreased, the advisory committee for that area was outraged:

> The thing which bothers us the most is that the [District Manager] made the cut against the advice of and contrary to the wishes of the Advisory Board. These men are all experienced stockmen—all are operators—they know the range capabilities—they are interested, even more than the Manager, in a long time operation. Certainly it was never the intention of Congress that this one bureaucrat should override the considered judgment of the cumulated experience of the members of the Advisory Board. *The Manager and his paid personnel should furnish the information and the board should fix the policy.*
>
> *If this is not the theory, it should be.* (Culhane, 1981, p. 90 [emphasis added])

Carpenter was not the only obstacle to range management by professionally trained land managers. Because the range management profession did not become well-organized until after World War II, the Grazing Service was not able to develop professional advocates, staff, and practices as rapidly as was the Forest Service. As a result, it was not until the 1950s that professional range managers began to infiltrate the organization (Culhane, 1981, p. 92).

In 1946, a dramatic change occurred in the Grazing Service. President Truman's Reorganization Plan Number 2 merged the GLO and Grazing Service to form the Bureau of Land Management (BLM). Marion Clawson was appointed as the BLM's first Director. In the stockmen's eyes, Clawson was "worse than a conservationist; he was an economist who believed the BLM could and should operate as a professional land management agency" (Culhane, p. 91). He was able to transform the BLM during his 7-year tenure (1946–1953) to a professional agency staffed with trained grazing managers and irrigation specialists (Culhane, 1981, pp. 88–91). By the early 1960s, Clawson's influence could be seen in the increasingly professional influence throughout the agency's hierarchy. Still, the influence of local interests (particularly in setting grazing fees and allocations) as opposed to professional judgment continues to distinguish Forest Service from BLM land management (Culhane, 1981).

A NEW ERA IN PUBLIC LAND MANAGEMENT

After World War II, the role of the U.S. Forest Service also changed dramatically. For the first 40 years following the Forest Service's establishment in 1905 to manage the forest reserves, its role had been primarily "custodial" (Culhane, 1981, p. 50; Clawson, 1971, p. 19–21). It managed the national forests essentially as reserves and sold little timber to private companies so as not to depress timber prices. The demand for public timber supplies increased markedly during the economic expansion following the war, however. The demand was further increased because private companies failed to manage their resources for a sustained yield, and, as a result, private inventories were becoming depleted (Culhane, 1981, p. 11; Clawson, 1971, p. 21).

With the postwar economy booming, private enterprise viewed the public land-management agencies to be obstructing needed growth. Efforts were made during the Eisenhower administration to shift responsibility for the public domain to private interests:

> . . . stockmen announced their goal of transferring public rangelands to private ownership, while the forest industry strongly advocated a prohibition against additions to the national forests and even a few schemes to transfer some federal timberlands into private ownership. (Culhane, 1981, p. 51)

Critically, these efforts were countered by those of environmental and preservation groups that had broader visions of the uses to which national forest lands should be dedicated:

> The Forest Service had been one of the darlings of the preservation-oriented conservationists because of its criticisms of private timber companies during the decades of private timber regulation conflicts. When the service indicated its plans to open the national forests to increased logging by those same companies, conservationists [sic] turned on the agency, opposing its timber policies and advocating increased recreational programs and preservation of wild and primitive areas. (Culhane, 1981, pp. 51–52)

Congress attempted on many different occasions to address this growing conflict between different users of the same public lands. It passed legislation clarifying its intent regarding what were legitimate land uses and how these lands should be managed. As is characteristic of the legislative process, however, these acts were necessarily broad in scope, giving only general guidance to land management agencies. Moving from broad policy statements to implementation in site-specific cases is no simple task. The land management agencies now must not only try to develop programs that satisfy the objectives of one particular act, but must, in so doing, accommodate numerous other policy objectives from often conflicting legislative mandates. Consistent with the paradigm, they do so in a manner that provides the greatest, sustainable return, considering the expanded agenda of legitimate uses.

Throughout this time, the land-management agencies fought policies or programs that would diminish administrative discretion (translated to efficient, professional planning) and allow single uses to dominate. The first major battle during this time was over the Wilderness Act of 1964 (16 U.S.C. 1131–1136). This act was viewed as a threat to the professional standards established by the Forest Service. It was not that the Forest Service did not believe in wilderness; indeed, it had administratively designated several wilderness and primitive areas over the years. The foresters believed in wilderness on their terms, when it would not interfere with the efficient use of commercially valuable timber resources, a very different conception from that of the preservationists. Professional foresters feared that a wilderness act would take the authority to designate these areas from the Forest Service and give it to Congress. Congress would then, undoubtedly, succumb to political pressures and be unable to make the professionally correct decisions. Thus, the gains made towards efficient and rational planning would be lost (Frome, 1974a).

While fighting the Wilderness Act, the Forest Service also advocated a multiple-use mandate, one that would acknowledge the many uses for which national forest lands should be managed, and thus free the agency from the demands of commercial interests:

> The act has frequently been criticized as an abdication of Congressional responsibility over the national forests because it is fairly vague and allows the service to make discretionary judgments among competing uses (for example, ". . . with consideration being given to the relative values of the various resources . . . "). It should really be viewed in the context of the 1950s—as a defense against extreme commodity user demands and as a codification of the Service's historic conservation mission to promote, as Pinchot put it, "the greatest good of the greatest number over the long run." (Culhane, 1981, p. 53)

With the Multiple-Use Sustained-Yield Act (16 U.S.C. 528–531) comfortably in hand, the Forest Service lightened its opposition to the Wilderness Act, which then passed 4 years later. The BLM similarly fought for a Multiple-Use mandate that it finally obtained in the Federal Land Policy and Management Act of 1976 (43 U.S.C. 1702).

The Wilderness Act was only the first legislative mandate to remove some Forest Service discretion over the uses to which the national forests would be dedicated. The Wild and Scenic Rivers Act, as well, gave Congress the final authority to designate rivers within this system (16 U.S.C. 1271). The National Environmental Policy Act (42 U.S.C. 4321), Endangered Species Act (16 U.S.C. 1361), and National Forest Management Act (16 U.S.C. 1600(note)) created additional steps in already established administrative processes and, in so doing, provided access for external input into and means for questioning agency decisions. More importantly, they each gave standing to appeal and sue to many noncommercial interest groups that before were unable to obtain "affected party" involvement. (This consequence, critical to the public land-management problem today, is discussed more thoroughly in Chapter 4.) Gradually, the administrative discretion enjoyed by public land agencies to make decisions based upon professional judgment began to erode. By the late 1970s, power had become fairly well distributed among the many users. And, more importantly, this power was consistently being exercised to undermine administrative decisions, "shared power" effectively becoming "shared impotence." (Of course, this consequence of a broadened environmental agenda is not unique to public land management. [See Altshuler and Curry, 1975.])

Professional land managers are now in a new era of public land management, one in which the conservation ideal and therefore the management paradigm are no longer sufficient to maintain their discretion. As new objectives—objectives contrary to the conservation ideal—are being legislatively mandated, the paradigm is falling short. It is not sufficient when land managers must make decisions in situations where noncommercial uses such as wilderness preservation, wildlife protection, and recreation are given the same status as the traditional commercial forestry and mineral uses. The professional scientific management paradigm fails because it is premised on use. Professional judgment reflects conservation values, not noncommercial and preservation values. And, although land managers continue to apply their long-established decision-making processes designed to achieve Pinchot's "greatest good of the greatest number in the long run," their efforts are not always successful.

Regardless, the conservation ideal remains today to be the objective sought by the public land agencies. Analyses of the Forest Service in 1960 and again in 1980, made the same observations about the Forest Service dedication to its mission: obtaining "the greatest good of the greatest number in the long run." In 1960, Herbert Kaufman commented that the competing demands confronting the U.S. Forest Service

> have given forest officers a sense of engagement in a crusade on behalf of the public interest. Their duties are elevated from routine forest management to safeguarding the economic, and perhaps even the military security of the nation. . . . They are placed squarely in the tradition of Gifford Pinchot. (pp. 223–224)

In 1981, Paul Culhane commented on

> the amazing consistency with which ranger interviewees mentioned "multiple-use" and Pinchot's "the greatest good of the greatest number in the long run" as the guiding principles in their work . . . and helps explain the Forest Service's tenacious commitment to the principles of progressive conservationism and multiple-use. (p. 69)

The same principles now guide the BLM. Although much power still rests with local advisory committees, the increasingly professional ori-

entation of the agency led to its 15-year battle for multiple-use legislation, finally acquired in 1976. In 1981, Culhane wrote that:

> The BLM has become almost indistinguishable from the [Forest] Service in
> one critical respect: as professionally trained resource managers, BLM of-
> ficers have a strong commitment to the principles of multiple-use manage-
> ment and progressive conservation. (p. 105)

Chapter 3 highlights the influence of the public-land-management
paradigm on the administrative decision-making process employed by
the U.S. Forest Service. It details the specific steps taken by the agency
in determining where and how oil and gas leasing and permitting may
occur, what timber will be harvested and how it will be harvested, and
how long range planning efforts under the National Forest Management
Act of 1976 are carried out. The chapter also discusses the potential
consequences of different agency actions, as well as the role of national-
forest-user groups in agency decision-making procedures.

The Process of Making National Forest Management Decisions

The controversy surrounding timber harvesting and oil and gas exploration and development on public lands is partly a function of the physical process by which they occur and partly a function of the administrative process by which decisions are made in managing the national forests. As this chapter describes, the Forest Service decision-making process is highly systematic, rational, and scientific. In fact, in some respects it provides a model that many agencies might do well to emulate. In order to understand why Forest Service decisions generate so much and so persistent a conflict, it is critical to understand not only what is at stake in these decisions, but also to understand the dynamics of the administrative decision-making process itself: who is involved and when, how problems are defined, how alternatives are developed and evaluated, and how final decisions are reached. The objective of this chapter is to probe these questions by sketching the decision-making process followed in managing timber and fuels resources, and in comprehensive forest planning. This chapter presents the decision-making process in theory; the following chapters illustrate it in practice, with an eye towards where and why the two differ and how reform might be achieved.

THE LEGAL CONTEXT OF NATIONAL FOREST MANAGEMENT

For the first half of the twentieth century, national forests provided barely 5% of the nation's timber supply. During the 16 years between

1950 and 1966, the agency harvested more than twice the amount of timber as was harvested in the preceding 45 years. This change was due primarily to private foresters who depleted the timber resources on their private forest lands following World War II (because of excessive cutting practices and poor or nonexistant reforestation efforts), and who, therefore, shifted their attention to national-forest timber supplies. The rationale behind the change resided in the fact that although national forests accounted for less than 20% of the country's commercial timberland, they contained more than half of the softwood sawtimber inventory (Wilkinson & Anderson, 1985, p. 138).

The agency was unprepared for the consequences of these increased timber-harvesting pressures. The consequences of various harvesting practices were not always well understood, and, unfortunately, were often revealed only after timber harvesting had occurred—often with tragic consequences. Agency clearcutting practices in West Virginia's Monongahella National Forest and Montana's Bitterroot National Forest led to Congressionally mandated reviews, and eventually new legislation that then focussed national attention on Forest Service timber-management practices. With passage of the Forest and Rangeland Renewable Resources Planning Act (RPA) in 1974, and the National Forest Management Act (NFMA) in 1976 (technically an amendment to RPA), formal timber guidelines for the Forest Service were codified. They marked a major transformation in the regulation of timber-harvesting practices and general planning of the national forests. As Charles Wilkinson, a public-land law expert, has written:

> Until the late 1970s, Forest Service timber planners operated with little statutory or regulatory direction. The 1897 Organic Act strictly limited the Forest Service's authority to sell timber, but neither national nor local agency planners observed those limitations. The Multiple-Use Sustained-Yield Act of 1960 [see Chapter 5] generally deferred to the agency's timber management policies. The Department of Agriculture's regulations for timber management planning during the 1960s and early 1970s occupied barely one column of the Code of Federal Regulations. The principle source of official direction for local timber planners was the *Forest Service Manual*. The NFMA converted many of the Manual's timber planning guidelines and procedures into statutory and regulatory requirements. (Wilkinson & Anderson, 1985, p. 119)

The system put into place by RPA and NFMA requires that the Forest Service complete an RPA Assessment every 5 years and undertake a comprehensive forest-planning program. Among other things, the mandated RPA Assessment must include:

> 1. An analysis of present and anticipated uses, demand for, and supply of the renewable resources of forest, range, and other associated lands with

consideration of, and an emphasis on, pertinent supply, demand, and price relationship trends; and,

2. An inventory of present and potential renewable resources and an evaluation of opportunities for improving their yield of tangible and intangible goods and services, together with estimates of investment costs and direct and indirect returns to the Federal Government. (36 CFR 219.4(b) (i))

Under NFMA planning, the information acquired through this RPA Assessment process is then applied to on-the-ground long-term forest planning. Using the RPA Assessment data, the Forest Service Chief establishes national objectives for a range of resource outputs from the national forest system. Specific resource outputs are then allocated to each of the nine forest regions. Each Regional Forester's Office is then responsible for developing the standards and guidelines to be followed by the national forests within their region in order to achieve their RPA Program objectives and resource output allocation. Finally, each national forest develops a comprehensive forest plan following these regional guidelines. Although the forest plans have a 50-year time horizon, they are to be revised every 10–15 years.

To ensure coordinated planning of all forest resources, Congress mandated that the Forest Service use an interdisciplinary approach to fulfilling its NFMA planning mandate. The agency uses interdisciplinary teams whose functions include:

1. Assessing the problems and resource use and development opportunities associated with providing goods and services;
2. Obtaining the public's views about possible decisions;
3. Implementing the planning coordination activities within the Forest Service and with local, State and other Federal agencies;
4. Developing a broad range of alternatives which identify the benefits and costs of land and resource management according to the planning process described in this subpart.
5. Developing the land and resource management plan and associated environmental impact statement required pursuant to the planning process;
6. Presenting to the responsible line officer an integrated perspective on land and resource management planning; and
7. Establishing the standards and requirements by which planning and management activities will be monitored and evaluated. (36 CFR 219.5(a))

This planning process must additionally be conducted in a manner satisfying the environmental assessment requirements of the National Environmental Policy Act of 1970 (described on pp. 60–62). It also includes at least 10 points at which public participation activities must occur.

Decision-making over where and how oil and gas exploration and development may occur on public lands is guided by the complex process established under the Mineral Leasing Act of 1920. In contrast to the statutory mandates governing timber harvesting and forest planning described above, in enacting the mineral leasing laws, Congress gave little thought to the environmental concerns that several decades later would dominate debate. As one analyst of the mineral leasing laws commented:

> Concern was focused on then popular matters such as the monopolistic and unfair competitive practices of the oil giants, Federal versus private ownership and development of natural resources, and conservation in the economical sense of prevention of waste. Accordingly, the legislative history depicts a law under which protection of the public interest is primarily intended to apply to economic and not environmental considerations. (Noble, 1982)

The Mineral Leasing Act and its amendments govern how private firms can acquire leases and permits to public land resources and then conduct exploration and development on them. Initially, regulations governed well-spacing and drilling practices to control the rate of production. Later amendments permitted regulations to limit production and to unitize pools. By requiring the unitization of single pools, the Secretary of the Interior gained authority to prevent the inefficient pumping of oil or gas by several lessees with leaseholds sitting atop the same pool. By requiring them to unitize—to combine their leases to form a single production operation—drilling would occur at a more efficient rate, thereby promoting greater resource recovery (Martin, 1982, pp. 368–369).

It was not until 1947 that one of the Mineral Leasing Act's provisions was construed more generally to protect other natural resources besides oil or gas. At that time Secretary of the Interior Julius Krug used his authority to require unitization limiting the number of wells drilled in order to protect wildlife and scenic resources in Jackson Hole. This "Krug Memorandum," and the Jackson Hole Stipulation enforcing it, have since been a standard attachment to all leases in the area designated by Krug (*Federal Register*, August 30, 1947, p. 5859). In other areas, the Bureau of Land Management or U.S. Forest Service generally condition leases in order to mitigate impacts to surface resources.

TIMBER HARVESTING AND MANAGEMENT

As outlined in the *Forest Service Manual*, in preparing a timber sale for bidding, agency officials must step through six "gates," each requiring specific outputs that ensure consistency and quality control in timber offerings (see Table 1).

TABLE 1. Timber Management Gates

Gate number	Gate title	Processes	Key activities
1	Position statement	Position statement development	Scoping, sale area selection, silvicultural exams, area logging/transportation analysis, economic feasibility, budget, and scheduling
2	Decision	Sale area design	Environmental and economic analysis, resource reviews, project transportation logging analysis, decision-making, prepare project activity plan, and silvicultural prescriptions
3	Timber sale preparation report	Sale plan implementation	Includes all field layout activities, document items for use in preparing appraisal, contract preparation, offering, and sale area improvement plan
4	Advertisement or notice	Final package preparation, review, appraisal, and offering	Appraisal, sample contract, advertisement, prospectus
5	Bid opening	Bid opening	Review bids, hold auction, tentative high bidder
6	Sale award	Sale award	Complete award activities

At Gate 1, a preliminary decision document is developed, providing an overview of a proposed sale to ensure an environmentally sound proposal as well as one that is timely and efficient. The District Ranger prepares this position statement after conducting an on-the-ground review of the sale area and studying those issues that are critical and unique to the proposed project. At Gate 2, a more intensive field investigation is undertaken in order to specifically design the proposed field layout for the area, ensuring that it can be done in an environmentally sound and economically efficient manner. At this point, if it is determined that these two driving objectives cannot be met in a particular area, the project is dropped. If it is determined that the project can be fulfilled in an environmentally sound and economically efficient manner, a project plan will then be completed. At Gate 2, transportation system needs are determined. Accomplishing the objectives of this gate involves an interdisciplinary process ensuring coordination and consistency with a forest's plan and other resource demands. The formal environmental assessment process required under the National Environmental Policy Act (and detailed on pp. 60–62) is completed at this time. Again, the District Ranger's office conducts these analyses.

Gate 3 commences the formal implementation of the sale plan developed in Gates 1 and 2. Necessary roads to complete the sale are designed to appropriate standards, given costs of construction and potential impacts on other national forest lands and resources. Financing for the roads is allocated in the most cost efficient manner. At this time, the economic analysis completed in Gate 2 is again reviewed to ensure that all silvicultural and logging systems, regeneration and slash disposal methods, and other sale activities impose only reasonable costs on the timber sale. At all points in preparing the sale, district staff must incorporate environmental coordination needs with cost minimizing methods and procedures for the sale. At Gates 2 and 3 the District Ranger is responsible for the following information outlined in the *Forest Service Manual* (Section 2403.12):

1. Determination of area on which cutting will be done.
2. Estimated quantity to be cut.
3. Conditions of sale, including:
   (1) cutting practice,
   (2) requirements for fire prevention,
   (3) slash disposal,
   (4) logging methods,
   (5) location and character of improvements, and
   (6) other coordination requirements and management decisions pertaining to the area.
4. Appraised value of the timber if a charge is made.

At Gate 4 the final sale package is completed, including a sample contract, bid form, prospectus of the sale and appraisal report. As outlined in Section 2403.2 of the *Forest Service Manual*, each timber sale contract must include practical requirements providing:

(a) Fire prevention and suppression measures;
(b) Protection of residual live timber, including established young growth;

(c) Satisfactory regeneration of timber as may be made necessary by harvesting operation;
(d) Prevention and control of soil erosion;
(e) Favorable conditions of water flow and quality;
(f) Complete utilization of the timber as may be attained with available technology;
(g) Reduction of hazards of destructive agencies [fire, insects, disease]; and,
(h) Minimal adverse effects on, or protection and enhancement of, other national forest resource uses, and other improvements.

Some standard timber contract provisions require specific timber cutting and removal methods of the timber purchaser in order to protect other resources. One standard clause, CT6.4—Conduct of Logging, requires that logging only be conducted with animals, trucks or rubber-tired tractors with a gross weight of no more than 13,000 pounds or a bulldozer of up to 15,000 pounds. Similarly, standard clause CT 8.2—Environmental Protection—allows the Forest Service to terminate the sale should excessive environmental damage occur. Gate 4 concludes with the formal offering of the sale through advertisement or notice of availability. Sales are announced at least 6 months before the scheduled bidding date, and sometimes as much as a year in advance.

At Gates 5 and 6, the timber sale is opened for bids, usually at oral auctions, and a contract is awarded for the timber. When the sale contract is signed and returned to the Forest Service, the high bidder must make a down payment of at least 10 percent of the total bid value. A performance bond must also be filed by the high bidder to ensure compliance with the contract terms. The purchaser usually has 1 to 3 years to complete the harvest and submit the remaining payment to the agency. At this point, the Forest Service formally closes the sale. Administering timber sale contracts requires the continued involvement of Forest Service personnel. Sale boundaries must be marked on-the-ground. Other approvals and agreements periodically must be made, including those for logging camp and road locations.

HARVESTING SYSTEMS

There are two main systems for harvesting timber. Under the first system—even-aged management—timber is harvested in a manner leaving stands of trees of a single age group. There are three methods used to attain even-aged management. The first method, *clearcutting*, involves removal of all the trees in a particular stand. The Forest Service limits clearcuts to 40 acres, although there are exceptions to this standard. New stands in the area are then regenerated either naturally, through artificial seeding, or by planting nursery-grown trees.

The second method, *shelterwood cutting*, involves removal of one-half to two-thirds of the trees in a stand, leaving the remainder for sheltering new seedlings. Once the new seedlings are well-established, the shelter trees are removed. The third system is *seed-tree cutting*. It involves removal of all but a few trees from each acre in a stand. These trees are too few to serve shelterwood purposes but are enough to provide seeds for a new stand of trees in the area. Again, once the new seedlings become established, the seed trees are removed.

In contrast, the second harvest system—uneven-aged management—involves *selection cutting*, the removal of select trees from a stand as they mature. These stands are entered every 10 to 20 years with either individual trees or small groups of trees marked for harvest. This system is referred to as uneven-aged management because the remaining stands of trees will always contain a mix of different-aged trees.

The National Forest Management Act contains specific standards governing timber harvesting practices in the national forests. Responding to the controversy surrounding the agency's clearcutting practices, Congress adopted what are called the "Church Guidelines." Named for the late Senator Frank Church who was then chairman of the Senate Committee on Interior and Insular Affairs' Subcommittee on Public Lands, these guidelines state that:

> Clear-cutting should not be used as a cutting method on Federal land areas where:
> a. Soil, slope or other watershed conditions are fragile and subject to major injury.
> b. There is no assurance that the area can be adequately restocked within five years after harvest.
> c. Aesthetic values outweigh other considerations.
> d. The method is preferred only because it will give the greatest dollar return or the greatest unit output.
> Clear-cutting should be used only where:
> a. It is determined to be silviculturally essential to accomplish the relevant forest management objectives.
> b. The size of clear-cut blocks, patches or strips are kept at the minimum necessary to accomplish silvicultural and other multiple-use forest management objectives.
> c. A multidisciplinary review has first been made of the potential environmental, biological, aesthetic, engineering and economic impacts on each sale area.
> d. Clear-cut blocks, patches or strips are, in all cases, shaped and blended as much as possible with the natural terrain.
> Federal timber sale contracts should contain requirements to assure that all possible measures are taken to minimize or avoid adverse environmental impacts of timber harvesting, even if such measures result in lower net returns to the Treasury. (16 U.S.C. Section 1604(g))

The message from these guidelines was clearly that timber harvesting, however it is conducted, must accommodate other national-forest resource values.

Timber can be logged, or cut, in a variety of ways. *Tractor logging* is by far the most frequently used and most economical method available, costing between $20–$30 per thousand board feet harvested. It involves use of large caterpillar or rubber-tire "skidders" that drag cut logs to the trucks for hauling to the mill. Because of their weight and frequent trips through a stand, tractors compact the soil and increase soil erosion in cut areas.

High-lead logging is a more expensive system ($30–$40 per thousand board feet) but does less soil damage than tractors. This system uses a cable, sometimes suspended from a high tower, to drag logs up to where trucks are located for hauling. Because logs more than 1000 feet away cannot be reached by this system, it often requires more roads than tractor logging requires.

Skyline logging, similar to high-lead logging, uses cables to pull logs to the trucks. It uses a strong cable stretched between two points, thereby allowing logs to be completely or partially suspended from the ground. It is useful in areas with very steep slopes and/or sensitive soils and can reach up to one mile from the landing point. Not surprisingly, it often costs twice the cost of tractor or high-lead logging.

Finally, *helicopter* or *balloon-logging* systems are used to carry logs off the ground for distances up to one-half mile. These systems can be used to protect highly sensitive soils, but because of their high costs are seldom used in the national forests (*The Citizens' Guide*, 1985, pp. 14–16).

TIMBER-HARVEST PLANNING

The timber-harvest-planning process falls generally under the NFMA planning program mentioned above. From the more general timber harvest goals established in the forest plan, a 5-year action plan is developed. This action plan maps out the general locations of those timber sales to be offered over the following 5-year period, indicates what harvesting methods should be used and which year each sale will be offered. Work on individual sales really begins 2 years before the sale is to be offered. Exact sale boundaries are identified, roads are delineated and harvest and logging methods are determined. An environmental assessment is completed that considers different alternatives. An interdisciplinary team is brought into the process, often including specialists from the Forest Supervisor's office, to develop and consider alternatives and determine exact conditions that should bound and guide any proposed harvesting. Some small sales are at times prepared at the discretion of the District Ranger. These district sales are not necessarily part of 5-year plans and can be sold a few months after being proposed.

IMPACTS OF TIMBER HARVESTING AND MANAGEMENT PRACTICES

Impacts upon other surface resources are unavoidable when timber harvesting takes place. Most impacts occur during the road-construction and timber-harvesting processes from the obvious disruption caused by the activity of loggers and their machinery. As described in a court ruling on a lawsuit brought against the Forest Service for its timber-harvesting practices, timber harvesting can have these impacts:

> soil erosion from road-building and timber harvesting, water quality and fishery degradation, air pollution from the use of fossil fuels and slash burning, adverse consequences from the use of chemical herbicides, pesticides, and fertilizers used in timber and range management, the decline in the number and types of wildlife and plant species . . . the destruction of solitude and beauty, and the diminution of wild and natural areas available for scientific study. (483 F. Supp. 465 (1980), p. 477)

Clearcutting can additionally increase fire hazard by reducing the fuel moisture content on the ground and can potentially increase the risk of insect infestation and disease since it creates stands of trees of the same species and age, making them more vulnerable to attack and epidemic. It can also destroy the recreational values of particular areas for up to 30 years (433 F. Supp. 1235 [1977]). Some timber harvesting impacts persist for years after the logging activity is completed. Additionally, the increased access to backcountry areas provided for hunters and poachers via logging roads can result in loss of viable habitat for some large game species (*Jackson Hole Guide*, December 13, 1984).

In addition to commercial timber harvesting, the Forest Service manages, with varying degrees of impact, existing stands of timber to maximize their timber productivity as well as to improve their wildlife-habitat resource capabilities. *Thinnings* are periodically conducted to cut away some trees in a stand so that those remaining will fare better and grow to larger diameters with the lessened competition for resources. Some thinnings are conducted by the agency itself because the trees removed have little commercial value. Commercial thinnings do occur when the trees to be removed are large enough to have value to a lumber company. The agency sometimes *fertilizes* stands of timber when low nutrients are potentially limiting forest growth. *Mortality salvages* are undertaken to remove dead or dying trees from a forest. *Regeneration* is used to control and guide the new growth in a cut area in a specific long-term management direction. Herbicides are often used to keep competitive brush or undesirable tree and plant species out of clearcut areas until the desired seedlings are well established there. Although selection cutting seldom requires herbicide use to control brush and has less visual impact than even-aged management systems, it does involve more frequent activity in particular stands and along associated roads (*The Citizens' Guide*, 1985, pp. 18–19).

OIL AND GAS EXPLORATION AND DEVELOPMENT

The mineral leasing process is complex because of the complexity of the exploration and development process itself. At the outset of exploration, it is impossible to know what oil and gas resources will be found and thus what further activity, if any, may follow. Consequently, the process by which decisions are made is incremental, with more extensive federal review and involvement occurring as more extensive development is proposed.

Oil and gas exploration and development occur in a succession of four interdependent stages: *preliminary investigation, exploratory drilling, development* and *production*, and finally, *abandonment*. Whether one stage even occurs is dependent upon the findings of that stage preceding it. Each subsequent phase is more involved and more costly than the one preceding it. Each stage is more threatening to the environment than that preceding it. And, each stage requires more extensive federal involvement than that preceding it.

No single decision determines where and how oil and gas exploration and development may occur. No single federal agency controls the decision-making process (see Table 2). Before an oil and gas company or individual can conduct preliminary surveys of the surface and subsurface indicators of an area's oil and gas potential, a *special-use prospecting permit* must be obtained from the surface land-management agency (U.S. Forest Service or Bureau of Land Management). Before any exploratory drilling may occur, an oil and gas company or individual must obtain the *lease* from the Bureau of Land Management to the tract being explored. If the tract to be explored is on acquired lands rather than public domain lands, the lease must be obtained from the U.S. Forest Service (or other federal agency with full responsibility for the lands). A lease confers the right to develop the oil and gas resources beneath a tract of land but does *not* permit exploratory drilling. Before any explora-

TABLE 2. The Four Stages of Oil and Gas Exploration and Development in the National Forest System and Required Federal Approvals at Each Stage

Stage	Federal requirements
Preliminary investigation	
Geophysical analysis	No approvals needed if no surface impact involved
Seismic testing	Special-Use Prospecting Permit required from USFS District Ranger
Exploratory drilling	Lease to the area must first be acquired from the BLM, subject to USFS Regional Forester's review and recommendation
	Permit to Drill must be acquired from the Minerals Management Service (MMS), subject to USFS Regional Forester's review and recommendation
Development and production	License from MMS subject to USFS Regional Forester review and recommendation
Abandonment	Bond released if USFS and MMS conditions satisfied upon agency review

tory drilling may occur, the lessee must obtain a *permit to drill* from the Minerals Management Service. Should exploratory drilling lead to a discovery, the lessee may not develop the field until a *license for development and production* is obtained from the Minerals Management Service.

Although all leases, drilling permits, and development and production licenses must be acquired from agencies within the Department of the Interior (Bureau of Land Management, or Minerals Management Service), DOI decisions are generally based on recommendations from the Forest Service when national forest lands are involved. The Regional Forester (in charge of one of the nine Forest Service regions) has responsibility for making all lease, permit, and license recommendations. However, this decision is always subject to initial review by the District Ranger and the Forest Supervisor for the national forest under application in that region. The only exception to this progression is for wilderness areas, primitive areas, recreation areas, and irrigation districts, all of which require the Forest Service Chief's review and final recommendation to the appropriate DOI agency (*Forest Service Manual*, Section 2822).

The remainder of this section describes the four oil and gas exploration and development stages, the associated environmental impacts, and the federal role during each stage.

STAGE I: PRELIMINARY INVESTIGATION

Oil and gas resources are elusive. They are hidden beneath the earth's surface and there is no way of telling precisely where they are hidden and in what quantities but by drilling. There are ways to determine where oil and gas might exist, however, in order to give some idea of where exploratory drilling should occur. This preliminary investigation—"defining a prospect"—can involve two steps: surface geophysical analysis and seismic testing. (See, generally, US Department of the Interior, Bureau of Land Management, 1981(b); U.S. Department of Agriculture, Forest Service, 1981; and, Noble, 1982.)

Surface geophysical analysis is the first step in defining a prospect. Using on-site surveys and aerial photographs of various exposed geological formations combined with information about exploration in nearby or similar areas, an initial prospect is defined and the probability that oil and gas might be found is narrowed. Surface geophysical analysis is conducted by an oil and gas company interested in obtaining a lease or already holding a lease and interested in developing its potential. As long as surface geophysical analysis involves no disruption of surface resources, it requires no permits or government review.

Seismic testing often follows surface geophysical analysis to give an oil and gas company more detailed information about the potential oil and gas resources of an area, and consequently whether or not expensive exploratory drilling should even be undertaken. In seismic testing, a straight line is mapped through an area with promising oil and gas potential. Shockwaves are then generated along this line. The shockwaves are used to map deep strata formations to indicate where potential oil and gas "traps" (oil or gas-bearing formations) may lie. The shockwaves echo back from the different geologic layers and are recorded by a series of sensitive geophones. These readings along a line are then combined to produce a profile of the subsurface geology (Sumner, 1982).

Seismic testing can be conducted in one of three ways. One involves heavy trucks carrying "thumping" or vibrating devices that pound the ground to generate shockwaves. The "thumping" method involves dropping a 3-ton steel slab several times along the predetermined test line. This steel slab is attached by chains to a crane on a special truck. The vibrator method employs four large trucks equipped with a vibrator pad that is about 4-feet square and is mounted between the front and rear wheels. These pads are lowered to the ground and then electronically triggered by the recorder truck. Like thumpers, the vibrators are then moved a short distance forward to continue testing along the line (U.S. Department of the Interior, Bureau of Land Management, 1981b, p. 18).

Using trucks, though, obviously requires roads or relatively flat and easily passable terrain. In roadless areas, shockwaves are generated by either surface or subsurface blasting with dynamite. Crews conducting this type of seismic testing in roadless areas use helicopters or, infrequently, horseback, for access (U.S. Department of Agriculture, Forest Service, 1981, C-1). Seismic crews are "leapfrogged" from one site to the next; while one crew is setting up a "shot" along a line, another is cleaning up from the last blasting. A "shot" consists of ten 5-pound sticks of dynamite suspended, on average, every 20 feet. Seismic crews average between 50 and 100 shots per day. For subsurface blasting, "shot holes" are drilled to between 50 and 200 feet deep (*High Country News*, January 9, 1981, p. 1).

Before an oil or gas company will undertake a seismic survey, costing between $18,000 per mile for surface blasting and $50,000 per mile for subsurface blasting, (U.S. Department of Agriculture, Forest Service, 1981, A-2), it is only logical that it would possess the leases to the area tested. On some occasions, a private seismic testing firm will conduct the testing and then sell their results to individual firms or industry associations. For example, the well-publicized conflict over oil and gas exploration in Montana's Bob Marshall Wilderness centered on a proposal by Consolidated Georex Geophysics (CGG) to conduct seismic testing there. CGG did not possess any leases to the area but was doing the exploratory work because several oil and gas companies were interested in obtaining leases there. (See Chapter 5 for more information about the Bob Marshall Wilderness dispute.) Because testing results are proprietary, crews may blast the same or slightly altered lines several different times but for different oil and gas companies (Sumner, 1982, p. 29). Usually, though, a firm will not make this expense without some assurance that it has control of the mineral rights should a promising prospect be defined. Therefore, leases are usually acquired before seismic testing is begun and long before any decision about eventual drilling can be made.

Unlike surface geophysical analysis, seismic testing does disturb surface resources and wildlife. Seismic blasting can start forest fires. With most seismic testing occurring during the summer or fall seasons when weather permits, there is conflict with other backcountry users. Frequent helicopter trips generate noise that disturbs wildlife and diminishes the backcountry experience for recreationists. Additionally, there is a risk that backcountry users or cattle ranchers will unknowingly cross shot lines when blasting is about to occur. Sometimes blasting contaminates groundwater supplies (Noble, 1982, pp. 120–123). These impacts, however, are all short term; little evidence of seismic testing remains 1 year later (Sumner, 1982, p. 29). A special use prospecting permit must be acquired from the surface land management agency before seismic testing may be conducted.

Federal Requirements: Special-Use Prospecting Permits

A special-use prospecting permit must be acquired from the surface land-management agency before surface disturbing preliminary investigation activities may be conducted. The USFS and BLM control prospecting on lands within their respective jurisdictions.

For national forest lands, this permit application must be filed with the USFS District Ranger responsible for the area to be surveyed. This application explains the planned survey methods, location, timing, and whatever project information is needed by the District Ranger to evaluate the proposal (*Forest Service Manual*, Section 2821.2). The District Ranger completes a brief environmental analysis to determine what conflicts with other land uses, if any, may arise. This environmental analysis also determines what stipulations should be placed on the permit to protect surface resources, wildlife, and other established or planned land uses. Informal consultation between the applicant and the District Ranger may occur to clarify or amend the proposed prospecting activity.

Various conditions may be placed on the prospecting permit to prevent or mitigate potential impacts to surface resources. These conditions vary depending upon the particular area and project plans. Should a proposed shot line run through a wildlife calving area or migration route, the surface land-management agency may hold back a permit until calving or migration seasons are over. Other measures often include avoiding all live streams by at least 100 feet; no activity during the Memorial Day, Independence Day, and Labor Day high recreation periods; no seismic activity during the grizzly bear denning period (October 15–April 30) or bald eagle nesting season (March 1–July 31); specific locations of helicopter landing sites and flight corridors; and, preventing blasting when extreme fire danger exists (U.S. Department of Agriculture, Forest Service, 1981, pp. C-1–C-4).

A prospecting permit does not confer any rights to the minerals discovered, nor does it give preference rights to the permittee in obtaining leases for the lands surveyed. The terms of the permit specify precautions to be taken by the permittee in protecting surface resources, preventing forest fires, and restoring the lands to their original state.

The permittee must post a performance bond to ensure that all stipulations will be met and reclamation undertaken once the testing is completed. A U.S. Forest Service representative periodically inspects the project to approve bond release and terminate the permit. The District Ranger is directed to approve the proposed project if it "does not create unacceptable impacts on the surface resources or unreasonably conflict with other uses." These permits are seldom denied but no prospecting permits are issued for lands expressly withdrawn from the operation of the mineral leasing laws (*Forest Service Manual*, Section 2821).

The Federal Leasing Process

As mentioned, an individual or oil and gas company is not required to possess leases before conducting preliminary geophysical exploration in an area. But, should this analysis indicate a structure worth exploring further, leases must be acquired before an exploratory drilling permit will be issued. Oil and gas leases are issued through three different systems, depending upon where the leasable tract of public land is located. When a tract of land has never before been leased and does not contain known oil or gas resources, it is leased noncompetitively *over-the-counter* to the first qualified applicant. When leases have been issued before but have since expired or been canceled, terminated, or relinquished, they are issued noncompetitively through a bimonthly lottery called the *simultaneous oil and gas leasing system*. Finally, when leases are for tracts of land within a *known geologic structure* (KGS) containing oil or gas, they are issued through a competitive bidding process. Each of these three processes is described below.

Over-the-Counter Leases. Leases issued for the first time in an area with unproven geologic reserves are known informally as "wildcat" leases. These leases are issued over-the-counter to the first qualified applicant. Any U.S. citizen can file an application for an oil and gas lease as long as he or she is not a minor and does not already hold leases for more than 246,080 acres in the state in which the applied-for lease is located (U.S. Department of the Interior, Bureau of Land Management, 1981a). The filing process is simple. All a prospective lessee must do is review the land plats for the BLM District or National Forest of interest in order to determine which tracts have not yet been leased. The lessee then chooses the tract(s) that interest him or her and files a one-page application for each (U.S. General Accounting Office, 1981, p. 5). A lease tract must, at minimum, comprise 640 acres; it cannot exceed 2560 (Clawson, 1971, p. 129). A $75 filing fee must accompany each application and, should the lease be granted, a $1-per-acre annual rental fee must be paid. (The Department of the Interior has proposed raising the rental fee to $3 per acre for the last 5 years of the lease term) (U.S. General Accounting Office, 1981, p. 5). The lease term is 10 years. It is nonrenewable.

Lease applications are always filed at the Bureau of Land Management (BLM) District Office for the state in which the lease tract is located. The BLM has full responsibility for leasing public domain lands for oil and gas resources under the Mineral Leasing Act, as amended. It is charged with evaluating each lease application and attaching any conditions to a lease that are deemed necessary to protect the "national interest" in oil and gas exploration and development. If the lease is for BLM lands, the application is reviewed in the same office as filed. In its review, the BLM ensures that the land has not already been leased or been withdrawn from the provisions of the mineral leasing laws. The BLM also confers with the Minerals Management Service to ensure that the tract is not within a known geological structure. If the lease is available, the BLM must then fulfill the requirements of the National Environmental Policy Act (described on pp. 60–62) before it may be issued. Usually a "finding of no significant impact" (FONSI) is made and no environmental impact statement undertaken. The lease is then issued with the appropriate stipulations and attachments.

When a lease application is filed for land under U.S. Forest Service jurisdiction, the BLM forwards the application on to the Regional Forester for his or her recommendation. The Regional Forester is responsible for making the final recommendation to the BLM, but it is usually made after review and evaluation by both the District Ranger and Forest Supervisor. The USFS ensures that the lease has not been issued or withdrawn. The agency usually satisfies NEPA's requirements by developing an environmental assessment and concluding with a FONSI. The USFS generally, but not always, posts a public notice of its intention to lease. Some national forest headquarters also send notices to groups and individuals on their mailing list, as well as to the local newspapers. There is seldom any consultation with the lease applicant. Public comment resulting from the posted notice may be considered in assigning stipulations to the lease. Upon completing its review, the Regional Forester submits a recommendation to the BLM. The BLM has final authority for issuing or denying a lease. In fact, though, few lease denials are ever recommended to the BLM and the BLM has traditionally adopted the USFS recommendation (Noble, 1982, p. 142). Only when the BLM decision is contrary to the USFS recommendation must the BLM complete an additional environmental assessment before rendering its decision (24 IBLA 12,14 (1976); 23 IBLA 102 (1975)).

All lands under the control of the federal government are public lands. Within this large category are two types of land that are treated somewhat differently by the mineral leasing laws. Public domain lands are those that have *always* been under the federal government's jurisdiction. Acquired lands are those lands that have been given to or purchased by a federal agency. With regards to mineral leasing, acquired lands are governed by the Acquired Lands Leasing Act of 1947 (30 U.S.C. 351-359 (1976)). Essentially, this act extends the provisions of the Mineral Leasing Act of 1920 to acquired lands, but gives the surface resource management agency *full* control over leasing and exploration and development decisions. Thus, when the BLM receives a lease application for National Forest acquired lands, this application is forwarded to the Regional Forester for consent (or rejection) rather than recommendation (*Forest Service Manual* Section 2822).

Stipulations govern what activities the lessee may undertake and vary depending upon where the tract is located. Some stipulations are standard and are attached to all leases issued. BLM Form 3109-3—Stipulation for Lands Under Jurisdiction of Department of Agriculture—is attached to all leases for national forest lands. This stipulation governs surface use and restoration by the lessee in conducting exploration and development activities (U.S. Department of Agriculture, Forest Service, 1980, Appendix 1). Other stipulations are specific to a particular BLM or USFS region or leasable tract. For example, "no surface occupancy" (NSO) stipulations are attached to leases in wilderness areas or areas with fragile surface resources. (Less than 1% of all leases have this NSO stipulation) (*Congressional Record,* May 20, 1980, S5646). Specific wildlife protection stipulations are attached to leases when important calving, denning, nesting, or migration areas might be affected by exploration or development activity. Some stipulations govern the time periods during which exploration or development may occur. Others govern specific areas within a leased tract that may or may not be developed because of especially sensitive surface resources (U.S. Department of Agriculture, Forest Service, 1981, Appendixes B, C). Some stipulations are conditioned on potential future events (i.e., a proposed wilderness designation). NTL-6—Notice to Lessees and Operators of Federal and Indian Onshore Oil and Gas Leases—is the U.S. Geological Survey's Conservation Division (now the Minerals Management Service) attachment to all leases issued. NTL-6 specifies the requirements that must be fulfilled by the lessee before exploratory drilling or field development and production may be undertaken. Its 15 pages describe how a preliminary environmental review and final environmental analysis should be conducted. It outlines the steps to be taken in applying for a permit to drill and provides guidelines for preparing surface use and operating plans to accompany an application for a permit to drill. It also specifies how wells should be abandoned and the surface area reclaimed (U.S. Department of Agriculture, Forest Service, 1980, Appendix 2).

The Simultaneous Oil and Gas Leasing System. Originally, when leases expired, were canceled, terminated, or relinquished, they would have been restored to the surface land-management agency (U.S. Forest Service or Bureau of Land Management) and become available for leasing, once again, over-the-counter. But in 1959, the *simultaneous oil and gas leasing system* was established to facilitate leasing of previously issued tracts of land. The BLM had encountered considerable interest in previously issued leases and had difficulty determining who, in fact, was the "*first* qualified applicant" when these leases were returned to its hands. Unable to control the altercations that often arose between those claiming to be first in line, the BLM developed a new system for issuing these leases. This system is commonly referred to as "the lottery" because all applicants' filing cards are placed in a bin and then drawn for each available tract (U.S. Department of the Interior, Bureau of Land Management, 1981a).

As with over-the-counter leases, the BLM informs the USFS when it is going to post previously leased tracts on national forest lands for the next lottery. The U.S. Forest Service can then supplement the listed leases with any additional stipulations deemed necessary to protect surface resources. The USFS again makes a public notice of its intent and conducts an environmental analysis (*Forest Service Manual* Section 2822.42).

The lottery is held bimonthly (January, March, May, July, September, November) on the first day of the month. The list of available tracts are posted in the Bureau of Land Management State Offices 1 month before the drawing. Applications can then be filed for any tracts (but no more than one application can be filed per tract) until the 15th of the month. Filing is a simple process. The BLM provides the necessary filing card and the only information necessary is the applicant's name, address and signature. A $75 filing fee must accompany each application. A drawing is held for those tracts receiving more than one application. For tracts of land in the energy-rich west, it is not uncommon to have several hundred applications per tract. In 1980, the BLM issued approximately 7,500 leases through the lottery after receiving almost 4 million applications for these leases (U.S. Department of the Interior, Bureau of Land Management, 1981a).

Known Geologic Structures. Leases for tracts of land within a known geologic structure (KGS) are issued competitively at the discretion of the BLM. The process followed is very similar to one used in offshore oil and gas leasing. The tracts to be leased are selected and then placed on the auction block. Tracts are not more than 640 acres in size. Leases are issued to the highest "responsible qualified bidder" (30 U.S.C. Section 226(6)). This analysis is centered on conflict over noncompetitive leases. Land management decision-making is proving most difficult and controversial for non–KGS leases; uncertainty is greater and, unlike KGS areas, exploratory drilling has frequently not occurred there previously. Therefore, proposed exploration in non–KGS areas is more likely to threaten other land uses or relatively pristine areas.

STAGE II: EXPLORATORY DRILLING

Should seismic testing indicate that a promising geologic formation does exist, exploratory drilling plans begin. Exploratory wells are commonly referred to as "wildcat" wells because of their unknown potential. Nationwide, about 1 in 16 wildcat wells produce significant amounts of oil or gas. However, only 1 in 140 wells produce enough to succeed financially (U.S. Department of the Interior, Bureau of Land Management, 1981b).

Each wildcat well requires a cleared and leveled 3- to 5-acre site. The site accommodates a drill pad, a 100-foot derrick, a 10,000-square-foot reserve pit for drilling muds, and the more general operating facilities such as generators, fuel and water storage tanks, trailers, pipe racks, toilets, and either a water well or surface water pump (U.S. Department of Agriculture, Forest Service, 1981, A-4; Noble, 1982, pp. 121–130). Wildcat wells are drilled to an average of 10,000 feet (U.S. Department of the Interior, Bureau of Land Management, 1981b).

Exploratory drilling requires that access roads into the wellsite be constructed or upgraded should they already exist. These are generally 14- to 20-feet-wide graded roads. There is growing interest in using helicopter rather than road access in cases where the terrain is difficult to pass or when special surface resources would be harmed (especially in wilderness areas). Helicopter access is more than three times as expensive as road access; costs average $160,000 per airlift mile for each well versus $50,000 per mile for roads (U.S. Department of Agriculture, Forest Service, 1981, A-3).

Exploratory drilling activities last from 1 to 2 years. Commonly, 2 or 3 wells will be drilled during this exploratory stage. Costs average $9.8 million for a dry well and $11 million if a discovery is made. Should helicopter access be used, these figures increase to $15 million for a dry hole and $18 million for a discovery (U.S. Department of Agriculture, Forest Service, 1981, A-3).

Given the extensive surface-resource effects, a permit to drill must be acquired from the Minerals Management Service before a lessee can begin constructing an exploratory well.

Federal Requirements: Applications for a Permit to Drill

A permit must be acquired before a lessee can conduct exploratory drilling. An application for a permit to drill (APD) details the lessee's plans and is submitted to the Minerals Management Service (MMS), a relatively new DOI agency. Established in February, 1982, the MMS has the same staff, offices, and responsibilities as the former U.S. Geological Survey's Conservation Division (*Oil and Gas Journal*, June 7, 1982, pp. 66–67). This APD indicates where the exploratory drilling will occur, how access roads will be located and developed, where and how mud-pits and other drillsite facilities will be constructed, and where additional wellsites may be located. Site reclamation plans, once exploratory drilling is completed, must be included in this APD (U.S. Department of the Interior, n.d.).

Environmental impacts associated with exploratory drilling are obviously dependent upon precisely where the exploratory well is to be located; a wellsite on flat desert terrain will pose different problems than one located in a high mountain meadow. Consequently, although it has full responsibility for decision-making, the MMS forwards the APD on to the surface-land-management agency (U.S. Forest Service or Bureau of Land Management) for its review and recommendation.

The USGS and USFS developed a detailed "cooperative agreement" in 1977 that lists the responsibilities of each agency in responding to an APD. In this agreement, it is made clear that the USGS (now the MMS) is solely responsible for issuing permits and is "the sole representative with respect to direct contact with the lessees and operators in matters related to oil and gas" (U.S. Department of Agriculture, Forest Service, 1981, Appendix D). Nonetheless, considerable consultation does occur between the lessee and operator and the USFS while project plans are being developed and exploratory drilling activity is being conducted. A preliminary environmental review occurs even before an operator's plans are finalized and the APD submitted. This review identifies potential conflicts with other land uses or resources, and impact mitigation steps that might avoid these conflicts. The purpose of this review is to assist the lessee and operator in developing project plans and directing initial surveying and staking activities before they occur.

Once the lessee and operator's project plans are completed, they are filed with the MMS and USFS in the formal APD. A field inspection with

MMS and USFS officials and the lessee's operator and contractor(s) occurs in approximately 7 working days. The proposed wellsite, access roads, and other surface-use areas are reviewed at that time. Specific environmental impact mitigation measures may be discussed during this trip and the operator's plan amended accordingly.

Within 10 days of the field inspection, the USFS must submit its recommendation to the MMS. MMS officials then complete an environmental assessment on the proposed drilling. Seldom is an environmental impact statement (EIS) deemed necessary for exploratory drilling projects. (The first EIS on an APD was completed in early 1982—U.S. Department of the Interior, Geological Survey, 1982.) Unless USFS and MMS officials disagree about the need for an EIS, the permit will be issued at this time. No more than 30 days should have transpired between APD receipt and permit issuance.

When the U.S. Forest Service receives the forwarded APD from the MMS, it usually posts a public notice of its review of the proposal. Public comment resulting from this notice may raise considerations in its review and proposed conditions on the eventual permit.

Additionally, the Environmental Protection Agency requires all oil and gas lessees and operators to prepare spill prevention control and countermeasure (SPCC) plans. Although the lessee or operator is not required to file this plan with the EPA, EPA officials may request it and, should it not be provided immediately, fine the operator. If a hazardous material spill occurs, the EPA will then review the SPCC plan (40 CFR 112). In addition to fulfilling this EPA requirement, operators must also comply with Department of Transportation and Interstate Commerce Commission requirements.

State and Local Government Requirements

State and local requirements of lessees and operators vary depending upon the state and locality. Most states require notification and a monthly report if a well proves productive. Some states have environmental protection requirements that must be fulfilled. Local and county governments become involved when access, zoning, or rights-of-way issues arise (U.S. Department of the Interior, Bureau of Land Management, 1981b, p. 14).

STAGE III: FIELD DEVELOPMENT AND PRODUCTION

The impacts associated with exploratory drilling are extended and compounded should a discovery be made and development and production be warranted. An average oil and gas field is 640 acres, with well-spacing of 40 acres for oil and 160 acres for gas. Generally, this implies that 4 gas wells or 16 oil wells, maximum, can be located in one field (U.S. Department of the Interior, Bureau of Land Management, 1981b, p. 30). Each wellsite again requires a 3- to 5-acre cleared and leveled drill pad with a 100-foot derrick, 100' by 100' fenced reserve pit, and the general operating facilities. Once the wells have been drilled, the derrick will be replaced with a system of 20' high "horsehead" or "grasshopper" lifts to pump the oil and gas. Field development requires "in-field" access roads, pipelines, and utility lines from wellsite to wellsite, and temporary housing and associated structures for field workers. Additionally, pipelines and transmission lines into the field must be constructed. Onsite oil and gas storage tanks are required. Eventually, injection wells will be constructed to promote secondary or tertiary recovery of oil and gas resources. If helicopter access is required, a staging area must be constructed outside the field as well as landing sites within the area. Roads must be maintained and snow removed during the winter season. In some cases, a "sweetening" plant must be constructed when "sour gas" (hydrogen sulfide) is mixed with the natural gas. The average life of a producing field is 30 years—the range is 15 to 50 years. The life of a specific field depends upon the size of the discovery (U.S. Department of Agriculture, Forest Service, 1981, A-4–A-6; Noble, 1982, pp. 124–127).

Production is an expensive undertaking. On average, the cost of developing a field is $2.5 million per well. Once developed, production costs include $3 million to drill each additional production well, $10,400 per mile annually to maintain access roads, $96,000 per well annually to operate the producing field, $15 million per mile for powerlines and $750,000 per mile for pipeline construction (U.S. Department of Agriculture, Forest Service, 1981, A-5).

Before a lessee can develop the field, a license must be acquired from the Minerals Management Service.

Federal Requirements: Licenses for Development and Production

If exploratory drilling leads to discovery of oil and gas resources in commercial quantities warranting production, the operator cannot simply proceed to develop the field. First, a license must be obtained from the MMS. The review and evaluation process for this license is similar to that for an APD. The operator submits an operating plan that details how field development will proceed, what construction activities will occur, and where and how reclamation will be completed. This plan is forwarded to the USFS for review and recommendation. Consultation between the MMS, USFS, and the operator will likely amend the operating plan to mitigate environmental impacts and avoid surface-resource conflicts if possible. The MMS then completes an environmental assessment and, frequently, an EIS with associated public hearings and involvement before rendering its final decision. As with exploratory drilling activities, the operator must post a performance bond before undertaking development and production. Both USFS and MMS officials will periodically inspect the operations to ensure that all conditions are being fulfilled and all stipulations adhered to (U.S. Department of the Interior, Bureau of Land Management, n.d.).

STAGE IV: ABANDONMENT

Abandonment begins immediately once production is completed. The well is plugged and capped. Generally, an above-surface pipe "monument" is required that lists location and name of well. This requirement can be waived by the MMS, particularly when surface-resource concerns warrant it (U.S. Department of the Interior, Bureau of Land Management, 1981b, p. 39). All equipment, utility lines, pipelines, powerlines, and field facilities are removed. The disturbed surface area is recontoured and revegetated as closely as possible to its original condition (Noble, 1982, pp. 127–130).

ENVIRONMENTAL ASSESSMENT PROCEDURES

As mandated by the National Environmental Policy Act of 1970, the Forest Service completes environmental assessments on all projects "significantly affecting the quality of the human environment" (42 U.S.C. 4332). Section 1900 of the *Forest Service Manual* details Forest Service environmental assessment procedures. Although a few agency actions, particularly administrative actions, are automatically excluded from the provisions of NEPA (refered to as "categorical exclusions") most decisions directly involving on-the-ground resources can be made only after following these procedures. Timber-harvesting, and oil- and gas-leasing and permitting decisions, in most cases, are reached after environmental assessments are completed.

There are two different levels of environmental analysis undertaken. These are distinguished by their relative comprehensiveness as well as by how much opportunity for public review and comment on the analysis and final decision is provided. In all cases, an environmental assessment (EA) is conducted. The EA involves analysis of a proposed project and alternatives to it, in order to determine what the likely environmental impacts of the action would be. Completing it usually involves the input of a variety of resource managers in the agency. Should the assessment lead to the conclusion that no significant impact will occur, then the responsible agency official files a FONSI and makes a decision. In contrast, should the conclusion be reached that significant impacts will result if the project goes forward, then the scope and depth of the environmental assessment is expanded in a more comprehensive EIS. More than just scope, however, the EIS process provides additional opportunities for public input to agency decision-making, particularly when the legislatively mandated draft EIS is distributed for review and comment by other government agencies and private groups and individuals. These comments are considered in finalizing the environmental analysis and reaching a decision.

The EIS, as required by NEPA, must analyze:

> (i) the environmental impact of the proposed action;
> (ii) any adverse environmental effects which cannot be avoided should the proposal be implemented;
> (iii) alternatives to the proposed action;
> (iv) the relationship between local short-term use of man's environment and the maintenance and enhancement of long-term productivity; and
> (v) any irreversible and irretrievable commitments of resources involved in the proposed action. (NEPA Section 102(2) (C))

The responsible agency official has considerable discretion in how and when various publics are involved in the EA process. The *Forest Service Manual* merely states that "early involvement by interested or affected agencies, organizations and individuals may be important to assure that significant issues, concerns, and opportunities are described" (Section 1951.2), and "public involvement is important in formalizing alternatives. The extent of involvement depends on the issues, concerns, opportunities involved, and the kind and magnitude of the decision" (Section 1951.5). Should a full EIS be warranted, the direction for public involvement becomes more specific: "When the action is such that an EIS is required . . . or is highly probable, there shall be an early and open process to facilitate free and open communication with the public" (Section 1951.2).

This participation begins at the initial "scoping" stage when the issues to be addressed in the EIS are defined. The agency must file a notice of its intent to prepare an EIS in the *Federal Register* and must contact affected and/or concerned parties. Thirty days later, a formal scoping meeting is held to set the scope of analysis. determine what issues are significant, and which alternatives should be evaluated in the EIS. Once the agency drafts its EIS it must be distributed for review and comment by interested individuals or groups and other governmental agencies. No less than 60 days is allotted to this review process. As defined in the *Forest Service Manual*, the agency is then required to "review, analyze, evaluate, and respond to substantive comments" (Section 1952.54) on this draft EIS.

THE APPEALS PROCESS

The Forest Service has a highly structured and strictly enforced decision-making hierarchy that steps through four levels in the agency. At the first level is the *chief* administrator for the agency, located in Washington D.C., and having final decision-making authority over all national forest matters. The USFS Chief is not a political appointee but is rather an individual who has usually climbed up the ranks of the agency and who might hold the position through two or more different presidential administrations. At the second level are nine *regional foresters*, each with responsibility over one of the nine Forest Service regions. They oversee the administration of all the national forests in their regions and provide guidance for implementing Washington Office policies and directions at the forest level. The third level is the national forest itself, with 146 *forest supervisors*, each managing 1 or 2 of the 155 national forests in the national forest system. Finally, each national forest is subdivided into districts, each managed by a *district ranger*. There are approximately 650 districts in the national forest system.

Decisions by any of these "line officers," even if, for example, the "decision" is merely a recommendation by the USFS to the BLM or MMS regarding an oil or gas lease or permit, may be appealed by any group or individual affected by that decision. For timber harvesting and forest planning there is a 45-day period after a final decision has been announced during which an appeal may be filed. For oil and gas leasing and permitting decisions there is a 30-day appeals period (*Forest Service Manual*, Sec. 2822.47). The "statement of reasons" supporting the appeal must state how the particular individual or group is affected by the decision and the specific complaint with how the decision was reached. It must also indicate what specific agency action or changes in the decision the applicant is requesting in the appeal. Sometimes an oral presentation by the appellant and other intervenors may be given before the Forest Service official hearing the appeal (36 CFR 211.18).

An appeal is always filed with the next superior official in the agency's hierarchy. For example, a timber sale decision made by a district ranger would be appealed to the forest supervisor. If the forest supervisor upholds the district ranger's decision, it can be appealed to the regional forester and, again, to the Forest Service Chief. Should an individual or group feel that their concerns have still not been addressed in any of these reviews, the Chief's decision can be further appealed to the Secretary of Agriculture.

For oil and gas leasing and permitting decisions the appeals process has several additional steps. A leasing recommendation is made by a regional forester. An appeal of this recommendation would, therefore, be filed with the USFS Chief in Washington, D.C. If the recommendation is upheld by the Chief, it can be further appealed to the Secretary of Agriculture. Should the Secretary of Agriculture again uphold the decision, the USFS recommendation is forwarded to the BLM district office. A group or individual can protest this recommendation to the BLM district officer. If the recommendation is accepted and a decision made accordingly, this decision can then be appealed to the BLM Director in Washington, D.C., and then to the Department of the Interior Board of Land Appeals and, finally, to the Secretary of the Interior. As is true with timber-harvesting or forest-plan appeals, if the individual or group is still not satisfied, a lawsuit can often be filed and the federal agencies taken to court. At least 2 years would be consumed in this process.

CONCLUSION

The resource development and decision-making processes described above appear, in theory, detailed and extensive but essentially straightforward. At each step in decision-making, the different land managers review the proposed activities to determine what decision should be made and what conditions should be placed on any activity that might occur. This administrative decision-making process is consistent with the land management paradigm described in Chapter 2; it assumes that the decisions to be made are amenable to scientific review and analysis. As the next three chapters illustrate, however, the process is very different in practice. It is extremely political and, in many cases unfortunately, ineffective. Even though analysis is exacting and all apparent bases are covered, decisions reached are frequently and effectively disputed. As Chapters 4–6 indicate, the problem posed by national forest management today has changed quite dramatically from that encountered when the agency was first established, as well as during its first 50 years. However, because the administrative decision-making process has not changed in concert, the agency's management efforts are at an impasse in many national forests.

CHAPTER 4

The Politics of National Forest Management

On paper, the decision-making processes described in Chapter 3 appear detailed but straightforward. In theory, they are rational processes involving professional land managers reviewing proposals, assessing impacts associated with these proposals, evaluating several alternatives, and, only then, rendering a decision. In making decisions, land managers consult with agency experts from several disciplines and frequently with individuals from industry or public interest groups. Such consultation is consistent with the professional, scientific land-management paradigm. It assumes that with sufficient information about a proposal a land manager will be able to make an appropriate decision; one that efficiently utilizes public land resources and, in so doing, satisfies the public's interest in land management.

In practice, the process plays out a much different story with an expanded cast of characters. Lessees, operators, and land managers are not the only voices heard when leasing and exploration decisions must be made. Agency interdisciplinary team members are not the only ones speaking when timber management plans and sale proposals are announced. Frequently, other groups become integrally involved in Forest Service decision-making. These groups raise additional, often conflicting concerns, and thereby considerably complicate the decision-making process.

Each Forest Service decision allocates national forest resources. Thus, some decisions benefit some user groups at the expense of others. Because there is a lot at stake in each decision, all affected interests inevitably try to influence decision-making. The result, as is seen in this chapter, is a very different and considerably more politicized process than that envisioned when the many different laws affecting national forest management were enacted.

THE MANY PUBLICS INVOLVED IN NATIONAL FOREST MANAGEMENT

Whether or not a group organizes and how actively it tries to influence national forest management depends upon what it has at stake in a particular decision. Because of the ongoing wilderness reviews and forest management planning processes, industry and environmental groups are especially well-organized and knowledgeable about the U.S. Forest Service and how it makes decisions. The Wilderness Society has published a two volume manual on both the issues to raise in a national forest plan appeal as well as how to appeal a national forest plan (1986a,b). Similarly, the Oregon Natural Resources Council has a two-volume manual describing how to monitor and influence national forest timber sales (Atkin, 1986). Cascade Holistic Economic Consultants, through their *Forest Planning* magazine, has published similar guides for those following national forest issues. The agency's wilderness review and forest planning processes have also generated considerable interest in and knowledge about the particular areas proposed for commercial development.

In November, 1981, the Wilderness Society was monitoring oil and gas lease proposals in 28 Wilderness Areas, 25 proposed wildernesses, 12 BLM Wilderness Study Areas and 16 USFS Further Planning Areas in the six Rocky Mountain States (Idaho, Montana, Wyoming, Utah, Colorado, and New Mexico) (Gottlieb, 1982, pp. 13–17). In October, 1984, the National Wildlife Federation was challenging the road-building proposals of 13 national forests in the Forest Service's Northern Region alone (*High Country News*, October 15, 1984). Additionally, the Wilderness Society has 18 economists, ecologists, lawyers, and other staff working full-time and 12 working half-time monitoring the forest planning process, analyzing forest plans as they are released, and preparing appeals and lawsuits on final plans. In hopes of influencing the direction the Forest Service's planning effort takes nationwide, The Wilderness Society is spending $4 million critiquing 14 key national forest plans (*High Country News*, August 5, 1985).

When proposals to either issue oil and gas leases, permit exploratory drilling, or put up specific areas for timber-sale bids are being considered by the USFS, environmental organizations voice their concerns and make recommendations to the USFS. They participate in whatever formal public hearings are held and monitor decision-making informally through communication with Forest Service staff involved in the analysis and decision-making. The involvement of environmental organizations is persistent, especially when decisions have national or regional significance because they will be precedent-setting, or will affect an area of particular scenic, ecological, or recreational importance. Environmental group involvement seldom ends when a decision is made should that decision run counter to what they perceive to be the appropriate outcome.

Different groups and individuals value national forest resources differently. Each group assesses the risks and the benefits associated with forest management decisions differently. As a result, rarely do these groups agree on what decision the Forest Service should make.

Environmentalists are concerned about protecting the scenic and ecological resources contained in the public lands. They fear that, in some cases, energy development and timber harvesting will destroy these resources forever. William Turnage, former director of The Wilderness Society, offers one explanation of the environmental view:

> In Europe there was a long cultural tradition; societies that have been in place for tens of centuries have their great monuments, their great cathedrals, the symbols of their civilization. America is a much newer nation, a nation of people who are deeply attached to nature. It's formed our character. And to us, our cathedrals, the monuments of our civilization, are the National Parks, the great Wilderness Areas, the wild rivers, the eagles of Alaska. Those are the things that make Americans different and special. And we've learned more than any other people in the world to take care of those things, and preserve those things. (December 9, 1981, Public Television Broadcast: "James Watt's Environment: Promised Land")

Brock Evans, vice-president of the National Audubon Society, concurs:

> Of course we need resources, of course we need minerals and energy and all those sorts of things . . . But, more important, more apropos, is the question of do we need it from these precise spots, these last little places remaining? ("James Watt's Environment", 1981)

Regardless of the commercial resource potential of some lands, environmentalists argue that development simply should not occur. To them, the benefits to existing and future generations of preserving these lands intact outweighs the opportunity costs of the timber and fuel resources foregone, regardless of how extensive they might be. Environmental group interest is not limited to wilderness areas, however. They are concerned as well about how commercial activity in exceptionally wild and scenic areas is regulated, particularly areas with important wildlife or recreational resources.

Like national environmental groups, oil and gas and timber industry associations closely monitor precedent-setting and policy-level decisions. They routinely participate in site-specific cases having a broad impact, as well as in federal policy development and legislative activity. Although industry associations do not themselves apply for oil and gas leases and drilling permits or make bids on Forest Service timber sales, they do have a stake in these decisions through their memberships. The Rocky Mountain Oil and Gas Association (RMOGA), for example, has 650 member oil and gas corporations in the Rocky Mountain region. Because its members file lease and permit applications and otherwise actively pursue oil and gas exploration and development on public as well as private lands, RMOGA files administrative appeals or lawsuits in those cases that will affect its membership generally (e.g., *RMOGA* v. *Andrus*, 500 F.Supp. 1338). Similarly, industry interest groups such as the Mountain States Legal Foundation (MSLF) and the Pacific Legal Foundation (PLF), advocate development interests in important cases, just as environmental interest groups support preservation objectives (e.g., *MSLF* v. *Andrus*, 499 F.Supp. 383; *PLF* v. *James Watt*, 529 F.Supp. 982). Representatives of the Western Timber Association have actively followed the development of national forest plans along the Pacific Coast. In frequent speeches before local chamber of commerces, county commissions, and other influential organizations, they have stressed the potential negative impact of timber-harvesting reductions and constraints on local economies. Their efforts on the West Coast as well as in Idaho have led to political pressures on the agency, delaying completion of forest plans there.

Predictably, industry groups approach oil and gas exploration and development differently from the environmentalists. Nonetheless, Kea Bardeen, attorney for the Mountain States Legal Foundation, does not believe MSLF's concerns to be in opposition to those of preservationists:

> The fact that we don't go around advocating wilderness protection all the time doesn't mean that we don't believe in Wilderness, that we would like to see it all leveled. . . . it's extremely important. But . . . wilderness values and mineral values are not necessarily incompatible. And that's because . . . [we] don't see wilderness as having to be totally pristine for all time. ("James Watt's Environment", 1981)

To preservationists, development of any type is not compatible with the Wilderness Act ideal of an area "untrammeled by man, where man is a visitor who does not remain" (16 U.S.C. 1131). They believe that wilderness, once developed, can never be truly restored to its wild state. This criterion has been amended, however, to encourage wilderness designation of previously logged or developed eastern national forests. (Frome, 1974a, Chapter 11; Healy, 1977) To industry interest groups, however, much commercial development is short-term and reversible. Perceiving the quality of wilderness differently, industry groups assess the costs and benefits of their activities differently. In contrast to preservationists, they conclude that the costs to society are much greater if oil and gas exploration or timber harvesting is prohibited. Thus, these groups argue that development, not preservation, best promotes the public interest.

Not all groups frame their arguments in terms of the "public interest." Some groups have more parochial interests at stake in these decisions. For example, outfitters protested exploratory drilling in Wyoming's Gros Ventre range because they feared it would scare away wildlife and, thereby, the hunters they guide through these mountains for a living (*Jackson Hole Guide*, November 17, 1977). Outfitters are concerned as well about the consequences of timber harvesting on their businesses. Timber harvesting and the associated road building have been blamed for the departure of one-third of the outfitters operating in the Bridger-Teton National Forest, translating into an annual loss in excess of $3 million to the local economy and the equivalent of 72 full-time jobs (*Jackson Hole Guide*, May 20, 1987). Ranchers who graze their cattle in the national forests argue against the seismic testing with dynamite that criss-crosses many western forests. This testing threatens both the cattle and the stockmen who unknowingly cross over shot lines when blasting is about to occur.

Now that industry interest in the resources contained in public lands is intensifying, proposals more frequently conflict with established users. Controversy is generated by proposals in heavily used backcountry areas, recreation areas, or areas deemed to be exceptionally scenic or wild. When the National Cooperative Refinery Association (NCRA) first proposed an exploratory well in Cache Creek Canyon outside Jackson, Wyoming, it would likely have appeared a reasonable proposal to an outsider. A dirt road already went up the Canyon and other development activities had occurred there in the past (U.S. Department of Agriculture, Forest Service, 1981). An exploratory oil well did not seem out of place. Jackson Hole residents, however, were outraged. They questioned the proposal on the grounds that it would diminish the quality of life in Jackson Hole, tax the town's streets and public services and detract from the area's exceptional tourism and recreation opportunities (*Jackson Hole Guide*, July 5, 1979). Teton County Commission Chairman Bill Ashley protested:

> To us it doesn't make any sense to screw up prime recreational areas with roads and other impacts of oil and gas exploration unless it's as a last resort. In this area, oil and gas and even mining and timbering are somewhat incompatible with the high priority given to recreation and wildlife. (*Jackson Hole Guide*, December 8, 1977, p. A11)

Ralph McMullen, executive director of the Jackson Hole Chamber of Commerce echoed a similar sentiment, justifying his opposition to the well: "People visit Jackson Hole first to see the area and second to enjoy it . . . the Chamber's purpose is to cater to these desires, not detract from them" (personal communication, Ralph McMullen, November, 1981). Jackson Hole residents valued Cache Creek Canyon as their "backyard;" a critical part of their character. To them, the costs of losing this amenity were far greater than any oil and gas that might be discovered there. NCRA, on the other hand, believed that the potential oil and gas resources there did warrant exploration and possible development and that, once completed, the area could be restored to Jackson Hole's standards.

THE PROCESS IN PRACTICE

Because of the many competing values at stake in national forest management, the seemingly technical and straightforward decision-making process outlined in Chapter 3 is, in reality, highly politicized. This marked contrast between theory and practice can only partly be explained by the stakes involved and because these many groups are organized in supporting different decision outcomes. In practice, three pathologies afflict the process and, moreover, politicize the process by giving a substantive basis for the claims of conflicting interests without providing a means for accommodating their concerns:

1. The process is *not sufficiently informative or convincing*. Forest Service analyses, no matter how thorough and seemingly objective, do not indicate what decision should be made; a "right" choice is elusive. Moreover, because no decision can be proven to be the correct one, the process is not convincing to those groups who perceive a different outcome to be more appropriate than that reached by the Forest Service.

2. The process is *divisive*. It separates different interest groups into adversarial camps and encourages strategic behavior among them. It provides no means for bridging the obvious chasm between them and hence only exacerbates the political conflict over the decision that must be made.

3. The process is *not decisive*. Even when the Forest Service ultimately makes a "decision," the "decision" rarely ends the controversy. On the contrary, the decision merely begins the next phase of the real decision-making process wherein groups resort to appeals, lawsuits, and pleas to Congress in hopes of influencing a "final" decision more supportive of their interests.

The remainder of this chapter illustrates these three pathologies and their consequences. Drawing from several controversial oil- and gas-leasing and permitting and timber-harvesting cases, this analysis pinpoints where the actual process diverges from theory and to what end.

Numerous oil and gas leases are involved in the cases highlighted or mentioned in this analysis. Seven hundred leases are at stake in Montana's Bob Marshall Wilderness (Findley, 1982, p. 311). Two hundred leases were involved in Wyoming's Palisades area (U.S. Department of Agriculture, Forest Service, 1980). Decisions on 135 lease applications have yet to be made for the Washakie Wilderness in Wyoming's Shoshone National Forest (U.S. Department of Agriculture, Forest Service, 1981). Covering 180,000 acres, 257 leases are still outstanding in California's Los Padres National Forest, some having been filed as long as 10 years ago. In Vermont, 130 lease applications are under consideration, covering the entire Green Mountain National Forest. In national forests along the east coast, 4800 leases were outstanding in February 1982 (*Boston Globe*, April 22, 1982).

Final decisions on these lease applications have been delayed for up to 10 years. These decisions have either been appealed by dissatisfied user groups, or the Forest Service has deferred decision-making on them out of concern for other surface resource values that might be harmed by oil and gas operations. Given that the Bureau of Land Management issues approximately 12,000 leases total for *all* public domain lands each year, these outstanding leases for the national forest system are not an insignificant concern for federal officials. In 1980, the late Senator Henry Jackson, chair of the Senate Committee on Energy and Natural Resources, wrote then Secretary of the Interior Cecil Andrus to find out what was causing the delays and how extensive they actually were. Andrus responded:

> Appeals of BLM State Office oil and gas decisions cause immense delays in lease issuance. There is at present a 6–9 month backlog of protests on appeal before the Interior Board of Land Appeals. In addition, all action toward lease issuance is suspended for 120 days following an IBLA decision in anticipation of an appeal to Federal Court. While some delay is unquestionably necessary in the event that a legitimate protest or appeal is filed, the present adjudicatory process is being abused. (*Congressional Record*, May 20, 1980, S5651)

Andrus went on to describe the extent of the delay, using just one of the nine BLM state offices as an example:

> 72 percent of the 7400 oil and gas lease applications backlogged for more than one year [in the Wyoming State Office] are tied up in other agencies, mostly in the Forest Service, and most of these as a result of the RARE II wilderness review process. (*Congressional Record*, May 20, 1980, S5650)

Andrus expressed concern for these delays and suggested to Senator Jackson that more staff and resources by allocated to the BLM leasing program.

1980 was an election year and Cecil Andrus was succeeded as Secretary of the Interior by James Watt. Watt, a Colorado attorney, had been president of the Mountain States Legal Foundation, an industry interest group advocating greater energy exploration on public lands. Similarly, M. Rupert Cutler, Assistant Secretary of Agriculture with responsibility for the Forest Service, was replaced by John Crowell, an attorney with Louisiana Pacific Corporation, a major timber company. Both Watt and Crowell immediately set out to rectify problems each attributed to the environmental sympathies of his predecessor. However, Watt encountered formidable obstacles to speeding up the leasing and permitting process. Crowell was similarly blocked in his attempts to increase timber harvesting and road building in the national forests. Their efforts further politicized these issues and made them front-page news across the country. Notwithstanding their efforts, leasing and permitting decisions and timber harvesting and road building proposals remain ensnarled in administrative appeals and Forest Service reviews. The problem cannot be remedied by the efforts of an Assistant Secretary of Agriculture or the Secretary of the Interior, even though these individuals, on paper, have final authority.

The administrative decision-making process is at the root of the current impasse in managing many national forest resources. The three pathologies that afflict the process—that it is not sufficiently informative or convincing, that it is divisive and is not decisive—make the decision-making task interminable. Neither the actions of powerful political actors nor the expertise of professional land managers can produce decisions that are accepted and supported by all the key groups involved. Power over these decisions has become distributed, and the agency's decision-making process has not been adjusted accordingly to include these stakeholders in decision-making in a constructive and meaningful manner. Therefore, it is fostering an adversarial system in which national forest management disputes are perpetuated, not resolved.

I. THE PROCESS IS NOT SUFFICIENTLY INFORMATIVE OR CONVINCING

As described in the last chapter, when the BLM or USGS forwards a lease or permit application to the Forest Service, Forest Service officials respond by preparing an environmental analysis (EA) of the proposal. When a District Ranger puts together a timber sale for the bidding process, an EA is similarly prepared. This analysis covers the specific aspects of the proposal, potential impacts on other surface resources in the forest, and finally, ways in which these impacts might be mitigated. One point of an EA is to determine whether or not a full environmental impact statement (EIS) should be prepared as required by the National Environmental Policy Act. In theory, again, this process appears straightforward. It mirrors the type of analysis the Forest Service conducts in almost all of its decisions. In practice, this analysis does not sufficiently inform decision-making; it is not obvious when undertaking an environmental assessment how extensive the analysis should be and when enough information has been obtained with which to make a decision.

For example, in mid-July, 1986, Wyoming's Bridger-Teton National Forest announced three new timber sales. One of the sales—the Minnie Holden—is located in an area where 20 sightings of the endangered Northern Rocky Mountain Gray Wolf have occurred. Six of these sightings have been either in or immediately adjacent to the timber-sale site. The Forest Service specifically designed the 7000-acre, 5.25- to 7.5-million-board-feet sale to avoid impacts on the gray wolf (*Jackson Hole Guide*, July, 22, 1986).

However, the Wyoming affiliate of the National Wildlife Federation was not convinced by the Forest Service's environmental analysis, or assured of the wolf's continued well-being. They appealed the sale, arguing that the agency needed to undertake a more comprehensive investigation before rendering a decision on whether or not the timber should be sold. According to a Wyoming Wildlife Federation representative:

> We are not convinced empirically that the wolf exists there, but there are enough indications from observations to merit a thorough search with a wolf authority who can fly for wolves and conduct a ground search just before winter dispersal when wolf howlings are easiest to detect.
>
> The Forest Service did do some searching, but it was not intensive enough. We want the Forest Service to be 100 percent sure. Under the Endangered Species Act, in addition to the environmental assessment of the area, the Forest Service should have had extensive contact with the U.S. Fish

and Wildlife Service. They've touched base, but it was fairly superficial. (*Jackson Hole Guide*, July 29, 1986, p. A3)

Even should the Forest Service agree to pursue this more indepth analysis, however, it will not be obvious upon completing their assessment what the agency should then decide to do; the assessment will not tell them definitively whether or not the risks to the gray wolf from timber harvesting in the Minnie Holden outweigh the value of the timber resource involved.

THE LIMITS OF TECHNICAL EXPERTISE

National forest management decisions are not made in an *ad hoc* manner. The extensive *Forest Service Manual* spells out in great detail the procedures to be followed by district rangers, forest supervisors, and regional foresters in analyzing a proposal and making final recommendations and decisions. Unfortunately, these guidelines call for considerable judgment on the part of Forest Service officials. The guidelines are necessarily broad and flexible because each decision involves different types of areas and resources.

In response to an oil and gas lease or permit proposal, for example, the federal government must either reject the application, accept it as submitted, or accept it subject to certain conditions. Forest Service officials reviewing a lease or permit application must decide whether or not to accept it, and, if so, what conditions, if any, should accompany the lease or permit. In making these decisions, the *Forest Service Manual* (Section 2822.42) lists 10 factors that must be considered:

1. Statutory authorities.
2. Existing and planned uses.
3. Dedications.
4. Impact on surface resources.
5. Damage to watershed.
6. Degree of surface disturbance and difficulty in restoration.
7. Special values, such as wilderness character, archaeological sites, cultural resources, and endangered wildlife habitat.
8. Term of the lease and probable nature of operations.
9. Economic considerations, such as relative values of minerals and surface resources and scarcity of and demand for minerals.
10. Range of alternatives available for operations and land uses and for environmental protection.

But, within statutory and administrative limits, *how much* damage to the watershed is permissible, or *what degree* of surface disturbance should be allowed, is usually up to the discretion of the field officer. Furthermore, this same subjective judgment must be exercised in deter-

mining "the relative values" of oil and gas resources and surface resources; judgment *must* be exercised because the oil and gas resources at
stake are of unknown quantity and type. Moreover, how a Forest Service field geologist is to determine "the scarcity of and demand for
minerals," and then consider these data in decision-making, is not prescribed. This question has been debated without resolution for decades
by experts throughout the world. As a result, should a Forest Service
official decide that an area's fuel resources are critically important, and
that the loss of surface resources is negligible relative to the value of
these fuel resources, another professional can easily argue to the
contrary.

Even a consideration as innocuous as "statutory authorities" requires judgment on the part of Forest Service officials and therefore
allows for criticism and opposition by those disagreeing or those who
are adversely affected by the decision. To a large extent, debate over
leasing in the Palisades area of Wyoming and Idaho centered on the
Forest Service's authority to enforce stipulations that it attached to the
leases. Similarly, debate still rages over the authority of the Forest Service and the U.S. Geological Survey to deny a drilling permit for environmental reasons in the proposed Gros Ventre Wilderness Area.

The *Forest Service Manual* (Section 2821.03(3)) guidelines describe
those instances where a prospecting permit or lease may be denied:

> A [prospecting] permit may be refused if the degree of disturbance will be
> excessive and result in unavoidable serious impacts on other resources.

Oil and gas leases may be denied when the Forest Service's environmental assessment indicates that oil and gas activity in a particular area
would:

> (1) seriously interfere with other resource values, (2) be incompatible with
> the purpose for which the area is being used or administered, or (3) perma-
> nently destroy or render useless the land for the purposes for which used or
> dedicated. . . . (or) when the value of the land, and its resources, for the
> purpose for which it is being used outweighs the foreseeable benefits that
> would be derived from extraction of the mineral resources, and the existing
> use cannot be adequately protected by stipulation. (*Forest Service Manual*,
> Section 2822.46)

Oil and gas leases may also be denied when an area has been withdrawn
from the mineral leasing laws. But, specifically withdrawing a particular
area in order to preclude leasing is discouraged:

> There should be relatively few requests for withdrawals from operation of
> the mineral leasing laws, because the land and surface resources ordinarily
> can be protected by proper stipulations, or because detrimental leasing can

be prevented by recommendations or refusal to consent to applications. (Section 2822.21)

The guidelines explicitly state what conditions warrant a withdrawal:

> Withdrawal may be requested if mineral leasing would (1) be incompatible with the purpose for which the land is dedicated, used, or reserved from use; (2) destroy or damage the values sought to be preserved; (3) hamper, restrict, or render useless the plans, programs, or functions for which the land has been utilized; (4) nullify major accomplishments and investments; or (5) create intolerable hazards or unjustified risk on lands having or planned for special purposes or programs, such as city watersheds, experimental forests, developed recreation areas, and archaeological sites. (Section 2822.21)

Again, as with land withdrawals, the guidelines make it clear that "wilderness designation shall not be sole justification for decisions against leasing, permitting or licensing" (Section 2822.46).

Forest Service field staff have broad discretion in decision-making under these guidelines. They must make a judgment about when a proposal "seriously interferes" with other land uses and surface resource "values." They must determine when existing uses "outweigh" the benefits of mineral extraction, or when disturbance might be classified "excessive" and its impacts "serious." Under these guidelines, any of a number of different decisions are possible on a single proposal; none is more correct than another. One district ranger might deem a proposal to be disastrous; another might view it to be inconsequential. A ranger living in a community adamantly opposed to a proposal might reflect these pressures in his or her decision; a ranger in a district far from the beaten path may be more easily influenced by the applicant's concerns.

The Forest Service acknowledges that judgment must be exercised in decision-making. In fact, the process provides for review by various agency officials of decisions and recommendations made by their subordinates. Each official is authorized to amend the decision or completely overrule it when his or her judgment indicates otherwise:

> After completion of the EA, or EIS if required, the Forest Service officer responsible for its completion will forward it, through channels, to the official responsible for the Forest Service decision. Each preparing and reviewing line officer will recommend an alternative, concur with or change previous recommendations, and provide an explanation for that position. Proper stipulations will be provided for each reasonable alternative, even though it is not recommended, in case the responsible Forest Service officer or the BLM does not concur with previous recommendations. (Section 2822.42)

In other words, the decisions to be made are admittedly nonobjective. None can be analytically proven correct. Just as these Forest Service line officers can exercise judgment in reviewing and adjusting a subordinate's decision, so too can those groups and individuals potentially affected by a decision.

Whereas agency officials do try within the bounds of their process to consider and incorporate the concerns of the wide range of public-land users affected by their decisions, the process does not convince these different land users either that the agency has tried or has successfully done so. For example, when the Bridger-Teton National Forest announced plans to put up the Fish Creek timber sale for bidding, Fred Kingwill, the forest's public information officer, presented it as a "noncontroversial" sale. The agency had incorporated several mitigating measures into the timber sale proposal in order to protect wildlife habitat, particularly that of the threatened grizzly bear. A combination of logging techniques were worked into the proposal and clearcuts were limited in size. Bill Knispeck, the forest's timber and fire officer, explained that:

> This sale doesn't involve roadless area questions and doesn't preclude any options in the new forest plan. You'd really have to work at it to find something to object to. (*Jackson Hole Guide,* August 8, 1985, p. A3)

The Sierra Club's Phil Hocker responded to the proposed sale with anger, however:

> I don't see how they can call a sale at the edge of situation one habitat "noncontroversial." I think they know their timber program isn't favored in Jackson and they're trying to justify it through belligerence. (p. A3)

Although the proposed timber sale encompassed some Situation One Grizzly Habitat, considered essential to the endangered bear's recovery, most logging was to be conducted outside this habitat. (Many wildlife biologists believe that no activities should occur in Situation One habitats until the bear's population has significantly improved.) Fred Kingwill justified the Forest Service's position that they felt accommodated all concerns:

> We recognize the importance of the bear as a national resource and its status as a threatened species. At the same time we believe other objectives [livestock grazing, recreation, timber harvesting] can be accomplished in these areas and that coexistence is still possible. (*Jackson Hole Guide,* April 5, 1984, p. B8)

The Forest Service believed that the timber sale would have no consequence to the endangered grizzly bear and was otherwise environmentally harmless because of the mitigating measures they would employ. Environmentalists, however, were not convinced and immediately questioned the proposal.

The decision-making *process* did nothing to convince those concerned about the grizzly bear that the proposed timber harvesting would not harm the bear and that the sale was appropriately structured. Instead, it took a step in the opposite direction by raising the ire and ensuring the opposition of environmental groups who perceived the process and decision to be "belligerent."

Professional expertise, as exercised in the existing timber management and administration process, does not always reveal which direction should be taken in decision-making. As a result, agency officials are consistently caught in the middle, trying to find an acceptable middle ground, but with little guidance on how to do so. In making a difficult decision to keep several timber sales on hold until the forest plan was finalized, former Bridger-Teton National Forest Supervisor Reid Jackson justified his decision by stating:

> The possibility that timber sales in these locations will close long-term land management options that should be considered in the current forest planning process, and that harvest as planned may possibly change in a significant manner the unroaded big game habitat and related recreation hunting and outfitter-related businesses has also led me to believe these decisions, when considered together, constitute a major federal action which significantly affects the quality of the human environment. Therefore, these decisions ought to be considered in an environmental impact statement addressing current public needs and desires. I have additionally determined that the appropriate format for this consideration to be the Forest Plan and associated EIS. (*Jackson Hole Guide*, September 19, 1985, pp. 1, 9)

In the face of the immediate and angry response of several timber dependent communities he justified his actions:

> We are concerned about providing resource supplies to the various dependent communities. We're doing the best we can to still keep other options open. I can certainly understand the concern expressed . . . because of the impact the decision might have on Fremont County residents. We also have a lot of concern from people who want us to leave the options open. (*Jackson Hole Guide*, October 3, 1985, pp. 1, 13)

Because of the wide range of different values involved in national forest management, the Forest Service is unable to objectively represent each one in decision-making. In this case, whereas the decision clearly pleased the environmental community, local outfitters, and recreation interests, it did not convince the lumbermill towns that their interests were adequately considered. The decision was immediately criticized and opposed.

ASSESSING THE "RELATIVE VALUES" OF RESOURCES

Although Forest Service officials must exercise judgment in making management decisions, they do not take this responsibility lightly or minimize the importance of each decision. Precisely because judgment is involved, they are thorough in their analyses, acquiring data, developing alternatives, assessing impacts and potential mitigation measures, and, only then, making a decision. Several different approaches have been used by Forest Service officials to assess the "relative values of minerals and surface resources." Some have used numerical rating schemes, while others have used less exacting relative rankings of alternatives.

Consider, for example, the Palisades oil and gas leasing decision. Before making their leasing recommendation for the Palisades area of Wyoming and Idaho, the USFS Supervisors in the Targhee and Bridger-Teton National Forests completed their environmental assessment on the proposed leasing. From their own analyses as well as from comments received in three public meetings, the Forest Service study team selected five alternatives and five decision criteria with which to evaluate these alternatives. The alternatives were:

1. deny all leases
2. defer a decision until a later time
3. lease all of the Palisades area
4. lease a portion of the area that was not environmentally sensitive (11%) and defer a decision on the remaining 49% [40% of the Palisades area had previously been leased]
5. lease the less sensitive 11% and lease the remaining 49% with no-surface-occupancy [NSO] allowed (U.S. Department of Agriculture, Forest Service, 1980, pp. 27–28)

The five "evaluation criteria" used were:

1. protection of wilderness values
2. identification of energy resources
3. compatibility with all natural resources
4. consideration of potential changes in the socioeconomic environment
5. compliance with Forest Service direction and authority (U.S. Department of Agriculture, Forest Service, 1980, p. 27)

To determine their preferred alternative, the Forest Service team rated each of the five alternatives against each of the five criteria using a numerical scoring system. Alternative 5 (49% NSO) received the highest score with 10 and Alternative 1 (deny all leases) was second highest with a score of 9 (U.S. Department of Agriculture, Forest Service, 1980, p. 46).

Assigning a particular numerical value to an alternative depended upon the particular evaluator's perspective. A Forest Service official obviously applies professional knowledge and experience in his or her ratings. But members of the Sierra Club, The Wilderness Society, and other environmental organizations assigned different values and thereby reached different conclusions. What the Forest Service argued protected wilderness values, the Sierra Club argued threatened these same values. In recommending leasing in the Palisades area, the Forest Service contended that oil and gas exploration was compatible with all natural resources. The Sierra Club argued that a road and drilling rig in a roadless area were hardly compatible with the existing wilderness and wildlife resources of the Palisades.

In selecting Alternative 5 as its "preferred alternative," the USFS found that it:

> is responsive to the key issues raised by the public since it balances the intense opposing concerns of environmental groups and the energy industry; is an action alternative. Refusing to lease or deferment of leasing would be unresponsive to National energy needs and would fail to comply with Forest Service policy and direction; and, allows the land surface management agency (Forest Service) more time to properly assess the adverse impacts on resource values and to formulate realistic mitigation measures to more effectively offset these impacts. . . . [and] to prepare for the administrative impact that may result should a significant resource be identified. (U.S. Department of Agriculture, Forest Service, 1980, pp. 46–47)

This finding assumed that "opposing concerns" would also view the decision as "responsive." Additionally, it assumed that the Forest Service retained the authority and the power to control future actions under the leases issued. Environmental organizations disagreed with both assumptions. They voiced their concerns at public meetings, in writing to the USFS EA team, and in person to USFS staff in the Targhee and Bridger-Teton National Forest. When their concerns had not been addressed to their satisfaction, they took the agency to court.

Much of the controversy over leasing in the Palisades area was created by uncertainty about whether or not development on the leaseholds would ever occur and, if it did, whether or not the stipulations would be enforceable. This would lead one to believe that uncertainty would be diminished and analysis greatly facilitated when a specific drilling proposal is submitted to the Forest Service. As seen in the Cache Creek/Little Granite Creek cases, though, the disputes remain and are perhaps heightened. Again, there often is no agreement on what the boundaries of analysis should be at the outset nor on what the conclusions of this analysis indicate should be decided.

The Bridger-Teton National Forest and U.S. Geological Survey jointly prepared an EIS on two exploratory wells in Cache Creek and Little Granite Creek outside Jackson, Wyoming. The draft document, released in August, 1981, painted a bleak picture of the development impacts in either area:

For Cache Creek, it concluded:

> Should field development occur in the Cache Creek watershed it would likely produce high impacts on recreation, wildlife (particularly elk), and local culture for many years (20 years or more). . . . The potential for contamination of surface water would be increased. . . . A major elk calving area would be eliminated. . . . Visual qualities would be severely compromised. . . . Noise from development drilling and vehicles, the road system, and landscape disruption would make the area unattractive to many forms of recreation. (U.S. Department of Agriculture, Forest Service, 1981, p. 3)

For Little Granite Creek, it concluded:

> The Little Granite Creek road alternative . . . would create moderate to high impacts on riparian vegetation, fisheries, wildlife, visual esthetics, and wilderness attributes of the Little Granite drainage. . . . Reclamation of the road would be particularly difficult because of the steep terrain and unstable slopes traversed . . . scars would persist for many years along the mile of road directly below the wellsite . . . Field development . . . would produce high impacts on the present wilderness character of the area for 20 to 30 years. (U.S. Department of Agriculture, Forest Service, 1981, pp. 3–4)

To determine what action should be taken given these impacts, the USFS/GS interagency task force analyzed two alternatives for each proposal: two different access road routes for Cache Creek and one access road and the use of helicopters rather than roads to reach the wellsite in Little Granite Creek. The draft EIS also discussed the implications of full field development, should it someday occur, for each proposal. These alternatives were evaluated in the draft EIS using nine criteria:

1. Determine the area's potential to contribute to the Nation's energy needs.
2. Recognize the lease rights of NCRA, Getty, and other members of the Bear Thrust Unit.
3. Provide for visual quality objectives of the area as defined in the Cache Creek-Bear Thrust mapping evaluation.
4. Protect water quality, aquatic habitat, and riparian zones, both onsite and in access corridors.
5. Minimize adverse effects on wildlife, basically by recognizing diverse habitat needs and protecting big-game populations.
6. Maintain traditional recreational use of Cache Creek drainage.
7. Minimize man's intrusion into the recommended Gros Ventre Wilderness in the exercise of mineral exploration and development activities.
8. Feasibility of reclaiming disturbed areas to natural conditions at the cessation of activities.
9. Minimize impacts on the community of Jackson (overload of community facilities, noise, hydrogen sulfide, and other safety factors). (U.S. Department of Agriculture, Forest Service, 1981, pp. III–1)

The task force supplemented this list of evaluation criteria with an additional objective that "alternatives must be practical, economically feasible, and provide a balance with respect to environmental protection and exploration" (U.S. Department of Agriculture, Forest Service, 1981, p. III–1).

The task force then studied the impacts associated with each proposal on 19 "elements": soils, air quality, noise, surface water, ground water, vegetation, wildlife, fisheries, recreation, wilderness, cultural resources, visual resources, population, local culture, economics, employment, housing, community services and contribution to nation's energy needs (U.S. Department of Agriculture, Forest Service, 1981, Chapter 5). These impacts were rated either "high," "moderate," or "low," using a dot system to visually compare alternatives. With this data in hand, each alternative was evaluated against the nine criteria using a rating scheme:

> Each Government scientist (FS and USGS) working on the EIS was asked to rate the magnitude of the impacts of alternatives for his specialty, according to these three levels. "Low" impact assumes very little change from status quo conditions; "moderate" implies some change to the extent that these changes would alter or destroy critical/key habitat or highly valued activities, produce intense community conflict, be obvious to anyone, or offensive to anyone, as the case may be . . . the ratings were thoroughly reviewed by experienced USGS or FS personnel to assure a sense of reasonableness and consistency. (U.S. Department of Agriculture, Forest Service, 1981, p. V–1)

As might have been predicted, the Forest Service analysis was immediately called into question. Phil Hocker of the Sierra Club criticized the task force's evaluation method:

> The dot-system used for evaluation of alternatives is too simplistic to be credible. Subjective judgements are hidden in the assignment of dots, which are not supported by the text of the dEIS, the supplemental studies, or fact. The bias of the authors of the dEIS has affected the construction of the naive charts shown. (October 19, 1981 letter to John Matis, USGS Task Force Leader, "Re: Comments on Cache Creek-Bear Thrust dEIS")

Because he had no faith in the process used to analyze alternatives, Hocker immediately questioned its conclusions. He criticized the cursory attention given to eventual full field development should oil or gas be found in significant quantities. He listed several alternatives that he thought were plausible but that were not discussed in the draft EIS. He also questioned whether or not the action satisfied the intent of the Wilderness Act, as the authors contended. He concluded:

> The draft EIS under review is "so inadequate as to preclude meaningful analysis," and a retraction and issuance of a new draft or multiple drafts is required (40 CFR Part 1502.9(a)). The scope of the draft is imprecise and appears to shift within the document. The draft omits important alternatives, and includes others which are inconsistent with policy positions taken in the study. The draft adopts a new Federal policy on lease administration and

uses this new policy to justify other positions within the study; however, the sweeping impacts of adopting such a policy are not studied, as is required. . . . Specific impacts are inaccurately portrayed or omitted. The conclusions of the dEIS are incorrect and insupportable. (CFR Part 1508.18(a))

Hocker was joined by several other environmental organizations in criticizing the EIS. Again, some suggested that there were other alternatives that should have been analyzed but were not. Others reached different conclusions using the same data acquired by the Forest Service-Geological Survey task force.

That judgment must be exercised in making these decisions is hardly an earth-shattering observation. Economists have long warned against the impossibility of trying to maximize a decision along several dimensions simultaneously. Policy analysis and planning evaluation literature similarly discusses the difficulty of making social choices when many objectives are desired (see Litchfield, Kettle, and Whitbread, 1975; Stokey and Zeckhauser, 1978). Forest Service officials acknowledge these shortcomings. But, they argue that Congress has mandated that they make decisions by considering and weighing numerous factors. While there may be no precise formula to indicate a correct decision at the end of analysis, Forest Service officials feel confident that their professional judgment leads to decisions that closely approximate "the public interest." Furthermore, they argue that, whereas their decisions might have shortcomings, who else could possibly make these decisions any better? They are still the agency with forest management expertise. Seeing no other way to make these subjective decisions, agency officials systematically and thoroughly study each proposal and, considering the information gathered in this effort, use their judgment to make a decision. But, because these analytical methods are flawed and, moreover, subject to dispute, they are almost always called into question by those who believe different factors should have been considered or weighed differently.

EVEN EXPERTS DISAGREE

Not only is there disagreement between the Forest Service and environmental and industry organizations about the type of information needed for decision-making and the appropriate conclusions to draw from each analysis, there is also disagreement among federal and state agencies with land-management responsibilities and expertise. When Montana's Bitterroot National Forest officials proposed a timber sale on 9400 acres along Tolan Creek, partly in order to improve wildlife habitat, officials with the Montana Fish, Wildlife, and Parks Department supported the National Wildlife Federation and Defenders of Wildlife in opposing it. According to John Firebaugh, Regional Wildlife Manager for the State Department:

> The Forest Service says the cut would benefit elk by providing more forage in the area. Our position is there's adequate forage there now. The limiting factor for elk there is security—a place to hide from hunters. (*Wall St. Journal*, April 18, 1986, p. 1)

In another case, whereas the Forest Service concluded that some development could occur without impact in the Washakie Wilderness in Wyoming (U.S. Department of Agriculture, Forest Service, 1981), the National Park Service concluded that this same activity "would destroy the wilderness values of this wild, remote and indescribably scenic area . . . [and] will be detrimental to Yellowstone National Park" (*The Washington Post*, September 7, 1981, pp. 1–2). When the USFS and USGS recommended that exploratory drilling occur in Cache Creek Canyon and Little Granite Creek outside Jackson, Wyoming, Roger Williams, EPA Region VIII Administrator, protested:

> Exploration is only the first step in what could become a major developmental process. While this action is significant in itself, the policy implications extend beyond just the proposed action. Important precedents will be set for future oil and gas EIS's.
>
> The 5–7 miles of road construction needed would drastically and irretrievably alter the area's wilderness characteristics. The Gros Ventre is the largest single de facto wilderness area in the lower 48 states. The Gros Ventre as wilderness provides important watershed protection. It provides important habitat for not only grizzly bears, but also eagles. Due to its generally steep slopes and unstable soils, the Gros Ventre will not fair well from a resource development standpoint. (August 26, 1980 letter to John Matis, USGS Task Force Leader)

Similarly, Idaho and Wyoming State Game and Fish Department officials expressed concern to the U.S. Forest Service about leasing in the Palisades area:

> The remoteness of the Palisades Further Planning Area has made it a haven for wildlife. Elk, and especially mountain goats, need protection from human disturbance. . . . The best solution to protect wildlife is no further leasing . . . our recommendation is no further leasing. (December 10, 1979 letter from Joseph C. Greenley, Director, Idaho Fish and Game Department, to David Jay, Targhee National Forest)

Nonetheless, the Forest Service concluded from its environmental assessment that "there will be no significant adverse effects . . . due to oil and gas lease issuance" (Sirmon, 1980, p. 1). The agency recommended to the BLM that the leases be issued.

Not only do experts disagree about the appropriate conclusions to draw from environmental analyses and hence what decisions should be reached, they additionally disagree about how much and what type of information is needed before a wise decision can be made. One such disagreement arose between the Forest Service and several other federal and state agencies over a proposed road and eventual timber harvesting in Idaho's Nez Perce National Forest. Forest Service officials argued that construction of the 14.2 miles of new road should be considered alone, with future timber harvests associated with it evaluated in future environmental assessments:

> Each timber sale is unique and the alternatives available are unique to the individual location and proposed action. The public will have ample opportunity to comment and influence the proposed action before any timber harvest decision is made. This sequential consideration of effects and cumulative impacts is not inconsistent with NEPA. As each new action is proposed, the cumulative impacts of preceding actions are assessed and considered with the effects of the preceding action. (Peterson, 1984, p. 12)

However, because the road provided access to an estimated 80 mmbf of timber and, furthermore, traversed the recovery corridor of the endangered Northern Rocky Mountain Gray Wolf, the agency's assessment was contested in court on the grounds that a cumulative environmental assessment was necessary and required under the National Environmental Policy Act. The Idaho Department of Fish and Game wrote to the agency:

> We . . . feel that an EIS is essential for a better description of the cumulative effects of the project on fish and wildlife resources. (Thomas *et al.*, 1984, p. 4)

The U.S. Fish and Wildlife Service concurred:

> Cumulative impacts of this and other proposed activities on wolf, wolf habitat, and wolf prey base should be considered. (p. 4)

Similarly, U.S. Environmental Protection Agency officials wrote:

> it seems only logical . . . planned activities within such proximity of one another—which will without question have significant impacts to the Salmon River—would be drawn up and evaluated simultaneously. This would result in a more realistic examination of cumulative impacts of the combined projects. (p. 4)

USFS officials are caught in the middle of an unquestionably difficult situation. There is no agreement between public-land users, the USFS, and other federal and state resource managers about the boundaries of analysis in decision-making, the specific alternatives to be analyzed, the meaning of *Forest Service Manual* directives, and the conclusions that ought to be drawn from the findings of analysis. Forest Service field officials wish they had more explicit formulas to follow in decision-making. Because they do not, they compensate by being as systematic and thorough as possible, so that all affected user groups can see that their interests have been considered. As has been shown above, not all the groups involved are convinced.

The analysis conducted by the Forest Service is critical to making an informed decision. This analysis cannot alone provide sufficient information with which to make a decision, however. Assumptions must be made in selecting alternatives and bounding the analysis, and judgment must be exercised in reaching conclusions from the analysis. These subjective aspects of decision-making cannot be subsumed within technical analysis. Although the process may satisfy Forest Service officials that all pertinent information has been considered and convince *them* that the decision reached is the most appropriate, it does not similarly satisfy or convince other affected groups or individuals. If the objective of this process is solely to reach a decision, then it is successful. However, if the objective is to be decisive—to make decisions that are accepted and supported rather than immediately contested and undermined—then the process fails.

Inevitably, many groups have preconceived notions about what an appropriate decision should be long before the analysis is completed. The decision-making process does not convince these groups either that their preconceived notions are wrong or that the decision reached is right. Therefore, it encourages criticism and, moreover, provides a substantive basis for this criticism. The process does not provide the opportunity for mutual inquiry to better understand the issues involved and the merit of a variety of different alternatives. Affected groups are not given an opportunity to amend, support, or reject their early notions. The process does not convince these groups that the critical assumptions and value judgments that in the end dictate which decision will be reached, are the most acceptable. It is unlikely that additional analysis would have convinced the Jackson Hole community that permitting a well up Cache Creek Canyon is the right thing to do. The values at stake simply do not lend themselves to the type of technical analysis conducted on this exploratory drilling proposal. But, should there be a mutually acceptable alternative (in the Cache Creek case or in the dozens of other cases in which the Forest Service similarly finds itself), the process is not structured to determine what that alternative might be. In this sense, the process is not as informative as it otherwise could be.

II. THE PROCESS IS DIVISIVE

When the National Cooperative Refinery Association (NCRA) filed an application for a permit to drill (APD) for its leasehold in Cache Creek Canyon outside Jackson, Wyoming, the initial reaction from the local community was cautious and concerned but not immediately negative. By the time NCRA withdrew its application 4 years later, the community was, literally, up-in-arms against drilling in Cache Creek. As one local resident exclaimed during the revelry following NCRA's announcement withdrawing its APD:

> We told them we were gonna harass the hell out of them. They were warned that there would be vandalism and all kinds of trouble. People carry guns up here, you know. (*The Baltimore Sun*, October, 1981)

The specifics of NCRA's proposal were only part of the reason the community's opposition grew to be so adamant; the administrative decision-making process provided no option but to respond as they did.

As structured, the national forest management decision-making process is divisive. It promotes distrust between parties, it encourages adversarial behavior, it leads to extreme position-taking and, ultimately, it ensures opposition to whatever decision is rendered. Although conflict over many national forest management proposals is inevitable, the process provides no mechanisms for anticipating this conflict and trying to resolve the differences among affected parties. As a result, the process exacerbates conflict and inhibits accommodation of the interests at stake.

THE PROCESS PROMOTES DISTRUST

The "decision-maker" in oil and gas leasing or permitting cases is, obviously, the federal land manager. As a result, the process is designed to generate information to inform his or her decision. The relevant information from a lease or permit applicant is contained in the official application. The federal official obtains additional information by requesting it directly from the applicant and by undertaking on-site inspections.

The public, on the other hand, obtains its information second-hand from the federal land manager. In many respects, the federal land manager becomes the applicant's agent, describing the proposal and its consequences in public announcements and at public meetings. The process provides no forum for direct communication between the applicant and the public. As a result, any hostility or opposition over a specific proposal, by design, is centered on the federal agencies. For example, when opposition to NCRA's well heightened, Ralph McMullen of the Jackson Hole Chamber of Commerce framed the battle as one of the community against the government: "people here think they can fight the federal government and they do and they win" (McMullen, November, 1981). Given this context, distrust of administrative decisions is frequently built into the process; it is difficult for public groups and individuals to divorce the proposal from the federal officials presenting it.

As seen in Chapter 3, the timber planning and management process is a very rational one, building in specificity over time as general timber goals give way to specific sale area proposals. Therefore, in many ways it makes little sense to substantively involve the public until the agency itself knows what it wants to propose. Fred Kingwill, public information officer for the Bridger-Teton National Forest, describes the rationale for the agency's process, one that he feels—and that on paper appears to—provide the agency with the flexibility it needs in developing harvest proposals, while at the same time soliciting public input:

> We aren't permitted to release drafts of environmental assessment documents until after a decision is made. The public then has 45 days to appeal. But we want to involve the public from the beginning, so we provide environmental analysis packages that contain maps and plans. That enables us to address public concerns from the beginning and, in the long run, make better decisions. (*Jackson Hole Guide*, June 13, 1985, p. A3)

Although this approach succeeds in the majority of cases, it nonetheless poses considerable problems for a significant minority. In these cases, affected groups feel that the agency's decisions have been made before they are consulted, and they feel their input is of little consequence.

One area in the Bridger-Teton National Forest that has attracted considerable attention because of its substantial timber supplies yet outstanding scenery, recreation opportunities, and wildlife habitat is the Mt. Leidy Highlands. The Sierra Club's Phil Hocker refers to the area as the last link in an "emerald chain" surrounding Jackson Hole and encompassing Grand Teton National Park and several wilderness and roadless areas. He argues that:

> The real natural resource of this region is not timber but it is beauty and recreation. And as we've learned in Jackson, that can be an economic resource as well as a spiritual one. (*Jackson Hole Guide*, June 27, 1985, p. A7)

In contrast, Fremont County, Wyoming, grapples with the state's highest unemployment rate and needs Bridger-Teton timber if its lumber mill is to remain open. All groups were actively lobbying the agency to either foreclose or expand timber-harvesting activity there.

In mid-1981, the Forest Service announced plans for a 400-acre, 7.3-mmbf timber sale in the Mt. Leidy Highlands' Jack Creek area. The agency had completed its analyses for the potential harvest and felt confident that the sale had been "carefully designed to meet wildlife, visual, and forestry concerns." (*High Country News*, September 4, 1981, p. 3) The sale was immediately contested however. The Wyoming Wildlife Federation argued that the road-building and timber harvesting would have a severe impact on the elk population, considered the largest in the contiguous United States. Others criticized the agency for failing to respond to local public opinion and expressed further distrust over the agency's claims that the logging roads would be reclaimed after the harvesting was complete. They pointed to other, similar roads that have not been reclaimed as promised and that were currently being used for recreational purposes. Senator Alan Simpson, (R-Wyo) said that "this is the harshest correspondence (the Wyoming delegation) has had on anything in two-and-a-half years." (*High Country News*, September 4, 1981, p. 3)

Three years later, in April 1984, Bridger-Teton officials were in the early, scoping stages of their Ditch Creek timber sale, also located in the Mt. Leidy Highlands. The first scoping statement made public was immediately criticized as being far too vague for meaningful public comment. Howie Wolke of Earth First! complained that:

> You can't expect to gather pertinent input from the public if you don't tell them what you're really planning to do. So I think they should start the whole planning process over from scratch. (*Jackson Hole Guide*, April 26, 1984, p. 1)

He and other members of the Jackson Hole environmental community were concerned about the area's elk and grizzly bear habitat, its significance as a roadless area, and its value to the recreation and tourism industry of Jackson Hole.

Jackson District Forester Mike Herth responded:

> I don't think we need to repeat the planning process because we've already discussed the project personally with many people. And as far as the specific timber sales are concerned, we described them in general terms because we still aren't sure exactly what we want to do.
>
> Our next step is to develop management alternatives ranging from roadless to varying degrees of development. As soon as that is accomplished we'll present them to the public and base our decision on a combination of concerns expressed at that time. (p.1)

However, as time would tell, this approach can backfire. Providing for more substantive public involvement later in the process precludes the early consideration and, moreover, accommodation of key concerns.

More importantly, it encourages an escalating distrust of the agency's deliberations. When a final proposal is eventually announced, this distrust turns to opposition, even when the agency feels that they have accommodated all key interests.

Two months later, as the agency's planning process progressed, "The Committee to Protect Wildlife Habitat" was established and collected 425 signatures on a petition calling for the Forest Service to completely abandon any timber-harvesting plans and road-building in Ditch Creek as well as in other sections of the Leidy Highlands. They wanted an 85,000-acre roadless sanctuary to be established in order to protect elk migratory routes and wanted to limit all logging to the winter months. Upon presenting the petition to the Forest Service the committee's chair commented:

> I'll be interested to see how the Forest Service tries to rationalize this project despite public opposition represented by our petition. In my opinion, it is more interested in meeting its timber-harvest target than managing that area for multiple uses. (*Jackson Hole Guide*, June 7, 1984, p. A7)

The Forest Service's response to the petition was in keeping with the philosophy underlying their decisionmaking process. Joe Kinsella, District Ranger on the Jackson District, received the petition with the comment that:

> Formal petitions are fairly common in this kind of situation. It's an important part of the democratic process and a good vehicle for communicating with government agencies.
> I think they're trying to be as objective as possible and avoid an antagonistic relationship with the Forest Service. I'm convinced this sort of thing is positive and we should do our best to keep the channels of communication open. (p. A7)

However, a year later the agency was only further ensnarled in the dispute as they proposed additional timber sales in the area.

In mid-1985, Forest Service officials put up timber in the Gunsight Pass area of the Mt. Leidy Highlands for bids. When approval of the timber sale was first announced, the Forest Supervisor's decision was met with anger and surprise. The local environmental community had been communicating with the agency about the sale and felt assured that "a decision was weeks away from being made" (*Jackson Hole Guide*, June 13, 1985). Len Carlman of the Alliance was "absolutely floored" upon hearing the sale was to be offered:

> I don't suspect malice or deceit by the Forest Service, but we're really blowing it when it comes to communication. (*Jackson Hole Guide*, June 13, 1985, p. A3)

Story Clark put it a bit more strongly to Forest Supervisor Reid Jackson:

> You are precluding options and, possibly more importantly, jeopardizing any remaining trust in the forest plan. (*Jackson Hole Guide*, July 18, 1985, p. A14)

Concerns about the elk calving grounds and migratory routes present in the sale area, its proximity to Situation One grizzly habitat and, more generally, the appropriateness of timber harvesting in the Highlands had encouraged a dialogue on the proposal between the local environmental community and the Forest Service from the outset. Phil Hocker reacted to the proposed sale, saying:

> This is just the first in a series of sales that will march down the Gros Ventre drainage and turn it into a timber drainage. We don't think that should be allowed to happen up there. (*Jackson Hole Guide*, June 13, 1985, p. A3)

Len Carlman did not mince his words:

> This sale doesn't look like wildlife recovery to me. I think it's just another nail in the coffin of the Leidy Highlands. (p. A3)

Upon realizing that public involvement in the environmental assessment process had been incomplete, Reid Jackson quickly reversed his decision, putting the proposed timber sale on hold. Reactions, predictably, flip-flopped when the postponement was announced. Whereas Phil Hocker expressed his support:

> Reid has made a bold, courageous move toward a better future for the forests. I take off my hat to him. (*Jackson Hole Guide*, August 6, 1985, p. A4)

and Len Carlman commented:

> I know it sounds trite, but this really is a case of democracy in action. I think it shows that Reid (Jackson) is listening to the public and taking appropriate action. Reid didn't have to do this, but I think it's an example of responsible management. (*Jackson Hole Guide*, July 30, 1985, p. 1)

the Louisiana Pacific Corporation and its employees and residents of Fremont County were angered. Bob Baker, Lousiana Pacific forester, charged that:

> This is an arbitrary decision based purely on political considerations. It raises questions as to the future viability of multiple use and challenges the concepts of sustained yield and even flow, which program forest resources over the long term to sustain healthy economies in local communities dependent on forest resources. (*Jackson Hole Guide*, August 1, 1985, p. A7)
>
> By placing the Gunsight Pass, Bull Creek-Harness Gulch and Moose-Gypsum Creek timber sales in a "hold" status the Forest Supervisor has substantially reduced the available timber base to Louisiana Pacific's Dubois, Wyoming, mill and the dependent communities.

> It has created instability and uncertainty in the forest products industry and is a threat to the economic stability of Dubois. (*Jackson Hole Guide*, September 19, 1985, p. 1)

Louisiana Pacific (L-P) immediately appealed the decision to Regional Forester Stan Tixier, charging that it violated the National Forest Management Act, National Environmental Policy Act, Wyoming Wilderness Act and the Multiple-Use Sustained-Yield Act. Waiting for release of the final forest plan would only jeopardize their Dubois mill's existence.

In turn, four conservation groups (the Jackson Hole Alliance, Greater Yellowstone Coalition, Sierra Club, and Wyoming Outdoor Council) intervened in L-P's appeal on behalf of the Forest Service. Phil Hocker explained their concern:

> The Forest Service is trying to think its way through some changes in attitude, and we understand that can't happen overnight. (*Jackson Hole Guide*, October 1, 1985, p. A6)

Regional Forester Stan Tixier upheld Jackson's decision but ordered further that comparable alternate sales be offered to compensate for the board feet held in Leidy Highlands pending completion of the Forest Plan. Regional and national environmental group representatives were surprised by Tixier's interim sales decision. Len Carlman of the Jackson Hole Alliance commented that:

> It's a clear indication that the regional forester is allowing a timber quota to drive the forest, and that of all multiple uses, he regards timber as the most important. We had all this talk about compromise, and now it's right back to business as usual. This is a dangerous precedent and an inherent contradiction. (*Jackson Hole Guide*, April 8, 1986, p. 1)

Similarly, Phil Hocker lamented:

> It looks like a classic case of you win the battle but lose the war. L-P got pretty much everything it wanted.
>
> This is not a well-considered decision by the Forest Service. It locks them into a position they've always denied they held. It makes them much, much more rigid, and, if anything, it's going to make compromise over the new forest plan much more difficult. (*Jackson Hole Guide*, April 8, 1986, p. 1)

The Jackson Hole Alliance immediately appealed Tixier's ruling to Forest Service Chief Max Peterson. (*Jackson Hole Guide*, May 6, 1986)

One of the reasons it may be difficult to rebuild trust in the Forest Service is the agency's tradition of transferring its employees from one forest to the next. Although this practice is not as prevalent as it once was (as the agency becomes more sensitive to the needs of two-career families and the personal as well as financial costs of such moves), it is still a factor, particularly with line officers. Any rapport and feeling of trust that a community and specific groups might develop with a particular district ranger or forest supervisor, for example, is not necessarily immediately transfered to that individual's successor. Moreover, plans that may be drawn up with a community understanding may be implemented differently by a new official unfamiliar with the history and underlying intent of the plan.

For example, Dan Heinz was the district ranger on the Leadville District in Colorado's San Isabel National Forest from 1974–1977. At that time there was considerable public support for including one section of the forest in a wilderness area. Heinz discussed the proposal with its advocates and convinced them that the area could be managed as backcountry, maintaining the essence of its wilderness qualities while still permitting some timber harvesting. The Upper Arkansas Planning Unit Plan and Final Environmental Impact Statement were then developed by Heinz and his staff, closely involving the public. The plan called for annually harvesting 250,000 board feet, creating 80 natural-looking openings to contribute to elk habitat and maintaining all future roads along existing corridors. Heinz felt that these provisions were a workable compromise, having the support of the local community and environmentalists.

Heinz was replaced in 1977 by a new district ranger, Gene Eide. Eide entered the scene as demand for the district's timber began skyrocketing. To meet local demand for firewood he allowed up to 1.6 million board feet to be cut, more than six times that contained in the plan. Additionally, to facilitate cutting the firewood, a new road was constructed outside any existing corridor. Clearcuts as well were larger than the 2-acre limit designated in the plan. Needless to say, a controversy erupted, with some local residents charging the Forest Service with a breach of public trust.

Heinz, as well, was disheartened to learn of these changes:

> A lot of people trusted me to put it into some type of backcountry management. I bitterly regret not having gone for wilderness . . . It would have been a shoo-in. I personally talked people out of it. . . . I didn't feel it was necessary. (The area) could still produce some wood and protect the values. (*High Country News*, August 6, 1984, p. 4)

Eide, however, justifies his actions:

> Any time conditions change you can change the planning. It's meant to be flexible. I think we're following pretty close. (p. 4)

Further, he dismisses Heinz's current criticisms as the product of different professional judgments:

> His heart and soul were in this plan. But in any large organization you have differences of opinion. We obviously have one. (p.4)

Heinz, though, feels that their differences are not rooted in differing silvicultural practices. To Heinz, the plan can certainly be changed, but not without involving the public that helped to develop it in the first place and who have placed their trust in the Forest Service to implement it as intended. Because no such public involvement occurred, Heinz believes that "a major violation of the public trust" resulted (p. 4).

It is certainly true that not *all* Forest Service officials make decisions in a manner that triggers distrust and opposition from interest groups. Behaviors do differ, though, and current management direction does not always give user groups sufficient confidence in future activity to encourage their acceptance of agency proposals at face value. Nor does the national forest management decision-making process provide sufficient guidance for line officers in building and maintaining trust and effective communication among forest users, particularly through the direct and, moreover, consistent involvement of these groups in making and then implementing particularly sensitive decisions.

THE PROCESS ENCOURAGES ADVERSARIAL BEHAVIOR

Few disputes seem simple and straightforward for the agency. Most seem to involve several issues and several parties and then escalate because the administrative decision-making process promotes distrust and divisiveness. It encourages groups to become adversaries, questioning each others' motives and actions.

Take, for example, the Forest Service's aspen-management program in Colorado. Aspen grow in clusters out of an underground clone, not by seeds. When the trees die, so does the clone and then conifers move in and take over the area. If the trees are cut, however, new sprouts are sent up by the still living clone. The agency wanted to circumvent this process and preserve aspen stands in Colorado by annually clearcutting 1% of the degenerate aspen stands in the Gunnison, Uncompaghre and Grand Mesa National Forests over a 50-year period. The cut aspen would be used to make matches, waferboard paneling and wood packing materials.

Stockmen protested the harvest, fearing that the thickets of sprouts would inhibit current cattle grazing. State tourism, recreation, and conservation groups were concerned about losing the aspen amenity as well as potential effects on the area's wildlife and other resources. In contrast, several local communities supported the plan because of the jobs and associated economic benefits it would bring.

Whereas opponents refered to the management plan as "The Conifer Conspiracy," the Forest Service justified it in these terms:

> We basically had to go ahead with it. Almost all of the area involved has got insect infestation and disease. The aspen need to be treated for range and wildlife improvements, and there are large areas where the conifer are invading the aspen stands. (*Rocky Mountain News*, August 1, 1984, p. 13)

The ensuing dispute centered mainly on how the Forest Service planned to implement its program and, moreover, how it had gone about developing and revising this program in the first place. Gretchen Nicholoff, representing the Western Slope Energy Resource Center, one of two groups appealing specific aspen sales, commented that:

> We've tried to communicate to the Forest Service that we weren't happy with their aspen management guidelines. I kind of feel like they have ignored or don't understand what we have been trying to talk to them about. (p. 13)

Similarly, Adam Poe of the State of Colorado Joint Review Process Office cut to the heart of the dispute when he said:

> I don't think anyone really doubts the need to cut some old aspen. There is no doubt that the majority of the acres that you cut will regenerate. The question is how the cuts are made, what size, where and when. (*Rocky Mountain News*, July 29, 1984, p. 12)

Unfortunately, the agency's process is not designed to answer these questions in a manner that builds trust and reaches solutions that address people's concerns to their satisfaction. Consequently, it promotes distrust.

In fact, the agency acknowledges the shortcomings of its decision-making process. An internal memo that was leaked to the press, admitted that the process it had used to make decisions about the aspen treatment program was flawed. The document contained statements such as:

> Sale program moving with the speed of a jet and the public understanding moving like a horse and buggy.
>
> Presence of Louisiana Pacific in Colorado has resulted in public questioning of our veracity.
>
> Forest Plans do not effectively address aspen management. (*High Country News*, October 15, 1984, p. 12)

The document further admitted that there were differences of opinion in the Forest Service about the long-range effects of aspen cutting. Whereas the Rocky Mountain Region justifies aspen cutting to halt spruce intrusion, the Southwest Region justifies the same treatment to encourage an area's takeover by conifers. It also implied that the new Regional

Forester, who stepped in after the aspen-treatment program debate had begun, was not totally in favor of it:

> Regional Forester may not subscribe to past/present aspen management direction. (p. 12)

It bemoaned the "re-active" stance imposed on the agency's Office of Information because of the debate that was raging in both national and grassroots environmental organizations, as well as in the media, particularly in "muckracking" columns by nationally syndicated columnist Jack Anderson.

As the Forest Service developed its aspen management program, Louisiana Pacific Corporation began constructing two $15 million aspen waferboard plants in the area. Collusion was immediately charged by opponents of the aspen cutting. Matt Mathes, Forest Service spokesperson, countered the charges, arguing that:

> It does look bad. It does look suspicious. But all Louisiana Pacific did was take advantage of the need to treat the acreage. (*Rocky Mountain News*, July 29, 1984, p. 12)

Furthermore, in a videotape presented at public meetings on the program, then Regional Forester Craig Rupp said:

> I see no basis for saying we were interested in aspens for commercial reasons. Louisiana Pacific is another tool in our tool box. [Given the need to harvest 4000 to 5000 acres of aspen a year] it was like a blessing when Louisiana Pacific walked in.
>
> To do nothing is unacceptable to me.
>
> When you think of the Rocky Mountains, you think of aspens, fall color tours, wildlife . . . (*High Country News*, February 6, 1984, p. 15)

The Montrose County (Colorado) Commissioners actively sought approval of the Forest Service's aspen cutting program. They flew to Washington D.C. (aboard an L-P plane) to present testimony favoring the program and threatened to prevent future use of the county jail by adjacent San Miguel County unless that county quit its efforts to get compensation from Louisiana Pacific for the damages its logging trucks bestow on local roads (*High Country News*, October 15, 1986). The *Montrose Press* refused to print a letter which it deemed critical of Louisiana Pacific (L-P) and the local radio station canceled the "Colorado Speaks" program after a speaker critical of L-P was interviewed. In an area beset with double-digit unemployment, support for L-P is rooted in residents' desire for the 140-plus jobs that the waferboard mills will bring to the area.

Others were wary about the speed with which the aspen clearcutting program was instituted and questioned the possible connections between former L-P counsel John Crowell's appointment as Assistant Secretary of Agriculture with responsibility for the Forest Service and L-P's arrival on the scene in Colorado. Chuck Worley, director of the Western Colorado Congress, suggested there was a "little hanky-panky going on":

> Louisiana-Pacific came into this area and built a $15 million factory without owning one stake of aspen. I have never seen a coal company so stupid or so reckless with its money, that it would go and spend $15 million without having a contract for coal. (*High Country News*, October 15, 1984, p. 13)

A meeting called by a western slope citizen's group, the West End People's Association, was designed to get accurate information on the project distributed and to open a dialogue between the different groups on the proposal, but it quickly disintegrated into name-calling. Fifty people attended the meeting. Montrose County Commissioners asked those questioning the proposal:

> What have you got against putting people to work? Our welfare and social services this year in Montrose are going to go over by about a hundred thousand dollars because we don't have people working. We thought it was a wonderful idea to bring in Louisiana Pacific and provide jobs, improve the economy, and put the aspen to beneficial use. (p. 13)

Worley responded:

> Why didn't you guys inquire into a company that pays higher wages? I get the impression that you guys will go for anything that looks like jobs regardless of what they pay or how they treat their people or anything else. (p. 13)

He suggested that a different company might pay higher wages and fulfill the aspen treatment program objectives "on the right scale so there will not be severe damage to other important segments of our economy" (p. 13).

The meeting then fell apart as name-calling began. Montrose County Commissioner Robert Corey claimed that Worley was arguing for a "wilderness concept" and was "discriminating against Louisiana Pacific and using it to try to stop aspen management. I don't think Louisiana Pacific's labor relations have anything to do with it" (p. 13).

The Forest Service finally brought a halt to the meeting. The agency's range and wildlife biologist denounced what he termed "half-face" charges that had been made throughout the meeting against the Forest Service:

> I think when you start right off the bat and call us liars, you set the wrong tone for the meeting. As far as I'm concerned, we're all wasting our time and we should go home. (p. 13)

The manner in which the agency had conducted its analysis and presented its proposal provided little option but for the groups to assume these adversarial positions. There was little effort to involve the different parties in a way that would educate them about the Forest Service's objectives and how it planned to achieve them and, furthermore, how it would convince all groups that their concerns were being addressed. Not surprisingly, when the agency's specific aspen sales began appearing on the auction block for bids, appeals of the decisions to go ahead with the sales began appearing on the Forest Supervisor and Regional Forester's desks.

THE PROCESS DISCOURAGES COOPERATIVE OR COLLABORATIVE BEHAVIOR

Precisely because the administrative decision-making process promotes distrust and divisiveness, it discourages any efforts to collaboratively reach solutions acceptable to all parties. Furthermore, the decision-making process often avoids issues underlying specific disputes, in favor of attention to more immediate or site-specific questions. Those issues at the heart of a dispute are often overstepped, thereby diminishing any incentive for affected parties to cooperate with the agency.

The debate over upgrading the Union Pass road in the Bridger-Teton National Forest provides an example of this difficulty. On the surface, upgrading the Union Pass road would not appear to be such an objectionable proposal. The 3.8-mile road already existed as a jeep road, albeit in a condition not suitable for logging trucks. Upgrading most of it entailed only filling potholes and grading. Only the first few hundred yards needed to be rerouted to provide a more gentle grade for the large logging trucks. Moreover, Louisiana Pacific had agreed to pay the $18,000 for the road upgrade and to restore the road once the harvesting was completed.

Why was the Union Pass road upgrade so hotly contested then? Three factors caused this decision to bounce from one Forest Service line officer to the next with threats of lawsuits all along the way. First, conservationists and the local tourism, recreation-based community felt that it would give Louisiana Pacific a toe-hold in the Upper Green River Drainage that would eventually lead to further logging and environmental degradation there. Second, these groups did not trust Louisiana Pacific's promises to restore the road. Third, there were conflicting messages coming from the Forest Service about both future use of the road and future harvesting activity in the Upper Green River Drainage.

In 1984, the Forest Service put up the Little Sheep Mountain timber sale for bids, the first in the area since the 1960s. More than 100 square miles in the Upper Green River Drainage were clearcut in the late 1960s, with what all agree were "horrendous" effects. The economics of the Little Sheep Mountain Sale indicated that it would likely go to a Pinedale Mill. Louisiana Pacific bid high, though, anticipating the Union Pass upgrade and, moreover, that this upgrade would then open up access to the timber in the Southern Bridger-Teton for their Dubois mill. According to Fred Kingwill, "It was a business risk (bidding on Little Sheep Mountain sale) on their part, and it still is" (*Jackson Hole Guide*, July 31, 1986). Louisiana Pacific has a lot at stake in the decision and is therefore strongly lobbying for the upgrade.

When Bridger-Teton National Forest Supervisor Reid Jackson approved the road upgrade, his decision was immediately appealed by environmental groups. Forest Spokesman Fred Kingwill predicted that whoever loses the appeal "undoubtedly will appeal [Regional Forester] Tixier's decision to Forest Service Chief Max Peterson." He admitted that the whole issue would likely lead to court in the end (*Jackson Hole Guide*, July 16, 1985). Kingwill noted how "highly symbolic" the issue had become by then and revealed that some opponents "have even threatened to sabotage industrial equipment" (*Jackson Hole Guide*, June 20, 1985, p. A13). One Bridger-Teton official described it as "one of those issues that haunts your dreams" (*Jackson Hole Guide*, January 30, 1986, pp. 1,A10).

As Jim Barlow, president of the Wyoming Outdoor Council, explains:

> Our principal concern, and the thing that's motivated us all along, is we don't want to see the same desertification of the Upper Green that has happened on the Dubois side. LP has so ravaged their side of the hill that the only way they can stay in operation is to come over and ravage ours.
>
> They change the land, they change the watershed, and where they have been, the effect is that nothing will grow again. (*Jackson Hole Guide*, October 29, 1985, p. A3)

Len Carlman, Director of the Jackson Hole Alliance, agrees:

> It's amazing that such a small project on the ground can have such a significant impact on the whole Upper Green River country.
>
> Once the road is there, it's there. Those things are difficult to undo, regardless of what the Forest Service decision says. The presence of that road would potentially open up more of the area for timber sales. (*Jackson Hole Guide*, October 29, 1985, p. A15)

The Regional Forester's office tried to facilitate negotiations between Louisiana Pacific and the road's opponents. The office had been thrown a hot potato with the appeal; what decision it should render was unclear. Bob Baker, Louisiana Pacific's head forester responded that "we're very much of the opinion that the public support is for the road." Louisiana Pacific had given several petitions to the Forest Service containing 1100 signatures expressing support for the project. However, Baker noted that "we certainly have reservations that some of the parties will be willing to negotiate" (*Jackson Hole Guide*, September 10, 1985, p. A7).

The first negotiation attempt failed. The road upgrade was not the issue that was really in contention, it was the fate of the Upper Green drainage. Therefore, conservationists were not going to agree to the road if it meant the beginning of more harvesting in the Upper Green. In turn, Louisiana Pacific had no incentive, in this forum, to agree to limited harvesting there. Both sides viewed the appeal, and a potential lawsuit, as setting a critical precedent. Because the Forest Service had not directly confronted either their responsibilities regarding the viability of local mills or their long-range vision for the Upper Green, there were no bounds within which a constructive, dispute-resolving dialogue could occur. With these fundamental questions unanswered, the underlying differences between the groups impeded any successful negotiations.

A few months later, the Regional Forester again gave the parties an opportunity to negotiate a solution to the road upgrade dispute. The Regional Forester announced that he would reveal the preferred alternative in the yet-to-be-released Bridger-Teton National Forest Final Plan, after a morning negotiating session. When the negotiations once again failed, he announced that the Forest Plan did not provide for upgrading the Union Pass road. Furthermore, it reserved timber sales in the Upper Green for mills smaller than Louisiana Pacific's and located closer to the sale areas (*Jackson Hole Guide*, March 4, 1986).

As Fred Kingwill explained:

> We were looking for the right mix and balance for users on that portion of the national forest. It was tough. Louisiana Pacific thought they had the improvement of the road down pat. It was difficult to take that away, but I think the decision will hold because that's where public opinion is.
>
> We're telling Louisiana Pacific that the Dubois mill will not be saved by the harvests on that portion of the forest.
>
> The future harvest is going to emphasize Douglas fir and spruce fir which are much larger logs and more economically advantageous to the smaller mills in the Pinedale area. (*Jackson Hole Guide*, March 4, 1986, p. 2)

Dick Rydeen, Louisiana Pacific operations manager, commented that:

> None of this comes as any real surprise. It makes life very tough for us. There's an excellent chance that we'll appeal the Forest Plan when it comes out. (*Jackson Hole Guide*, March 4, 1986, p. 1)

After announcing the plan's preferred alternative, the Regional Forester's office gave the parties one last chance for a negotiated solution. The Deputy Regional Forester apparently believed that knowing the preferred alternative would encourage the groups to come to some agreement. Because future timber sales in the area were not suited to Louisiana Pacific's operations, it seemed logical that everyone would agree that Louisiana Pacific could use the Union Pass road for the Little Sheep Mountain sale and then reclaim it because there would be no future need for it. However, the Regional Forester's announcement was of the plan's *proposed* preferred alternative; the plan could still be appealed and altered, leading to a very different final outcome. Because Louisiana Pacific had no incentive to agree to a stay of road upgrading while the plan was being finalized, no agreement was reached (*Jackson Hole Guide*, February 25, 1986).

These negotiations failed also because the parties involved did not trust one another. Conservationists claimed that Louisiana Pacific had "a hidden agenda" to haul more timber than just that from the Little Sheep Mountain sale over the Union Pass road:

> You know they want to haul additional timber, and you know they intend to keep that road open just as long as they can. If you're just looking at three million board feet of timber, it isn't worth all this effort for Louisiana Pacific. They've got bigger things in mind. (*Jackson Hole Guide*, February 20, 1986, p. A10)

Furthermore, they questioned the Forest Service's ability to implement the final decision as intended. According to John Barlow, Wyoming Outdoor Council president:

> We don't have any faith in the Forest Service's ability to close that road or to control Louisiana Pacific. (*Jackson Hole Guide*, October 29, 1985, p. A3)

Additionally, he claimed that:

> Louisiana Pacific's intention is to delay the plan as long as possible so they can use the road for as long as possible. (*Jackson Hole Guide*, June 12, 1986, p. A3)

The decision-making process did little to break down these barriers built of distrust. It provided no mechanisms for binding the groups to whatever agreement might be reached or by assuring the validity of different parties' claims. Consequently, any efforts at collaboration were fruitless. The process provided no incentives for more cooperative behavior; all incentives were for each group to continue battling in successive avenues in hope of prevailing in the end. As a result, the process precluded finding a mutually acceptable outcome, should one have been possible.

THE PROCESS ENCOURAGES STRATEGIC BEHAVIOR

Not only does the process deepen the chasm between the traditional adversaries, it also encourages these groups to behave strategically. Because the process puts potentially affected groups on the defensive and prompts distrust, it encourages all groups to seek other means of protecting their interests; it encourages them to seek other avenues through which their power to influence the final outcome might be greater.

For example, when Reid Jackson announced that the three Leidy Highland timber sales would be postponed pending completion of the Forest Plan, residents of Dubois felt compelled to take action outside the agency's decision-making process. The manager of the local radio station took time on-the-air to encourage their response:

> Perhaps we have no direct control over Forest Service decisions as it places three timber sales in the Upper Gros Ventre and Upper Green River areas on "hold" status until it issues a new forest plan in 1986. But we do have the capability of screaming loudly, writing lots of letters, and making a heap of phone calls to public officials in Washington and Cheyenne.
>
> Fremont County residents do have an opportunity to vote, with letters, phone calls and a lot of noise, as environmentalists and well-intended federal decision-makers prepare to rip out another chunk of flesh from our county's carcass. (*Jackson Hole Guide*, October 3, 1985, p. 1)

Over time, the various interests with a stake in the agency's decision found themselves in different camps, with the agency's process encouraging their combative stances.

When the USFS and USGS decided that a no-drill decision for Cache Creek was beyond their statutory authority, the Jackson Hole community concluded that their concerns could not be adequately addressed in the administrative process. The Jackson Hole Alliance came to the conclusion that the EIS would be "little more than a justification for the project . . . There is no administrative remedy now that we have reached this stage of development" (Clark, 1981). The Alliance was "appalled that leasing had occurred without public input" and now that an APD is filed "the decisions have already been made and there is nothing in the administrative process that the community can do about it" (Clark, 1981) In their mind, involvement in the administrative process had become "meaningless" (Clark, 1981).

When the administrative process failed them, the community moved into a forum where they felt they had more influence: Congress. Their approach was to encourage Congressional action to exchange or buy back NCRA's leases. They were convinced that they would be successful. Ralph McMullen, the Chamber of Commerce's Executive Director, explained why:

> We have people who know what to do and whom to call . . . The people who move here often are powerful, wealthy and influential. The locals know where to go and our tentacles reach far. (*The Baltimore Sun*, October, 1981)

McMullen was convinced that the Congressional delegation would be much more responsive to the Chamber's concerns than was the USFS or USGS:

> The Forest Service and Geological Survey are bureaucracies and so they are not responsive to the electorate . . . but the Congressional delegation is republican and so are most of the Chamber's members . . . they listen to us. (Interview with Ralph McMullen, November, 1981)

The Chamber and the Alliance both felt they had a better chance to influence decision-making through Congressional channels than through administrative channels.

This strategy was encouraged by a seemingly favorable response from their Congressional delegation. Given the Jackson Hole community's united and forceful opposition to the Cache Creek well, Wyoming's three-member Congressional delegation gave its support to the community. Senator Alan Simpson reported that "all of us have grave reservations about drilling in that area. I haven't found anyone in the county who favors this, so we will pursue everything we can" (*Jackson Hole News*, August 26, 1981, p. 1). Senator Malcolm Wallop declared that "we can generally agree that it ought not happen. The delegation will try to find any means available to it to prohibit the drilling in Cache Creek" (*Jackson Hole Guide*, September 3, 1981, p. A7). The delegation assured Jackson Hole residents that they "don't believe oil and gas drilling will ever be a reality in Cache Creek Canyon" (*Jackson Hole Guide*, September 3, 1981, p. A7). With such an encouraging response, there seemed little reason for the community to pursue the administrative process.

When opposition to drilling in Cache Creek Canyon flowed over to Getty Oil's proposed Little Granite Creek well, Getty officials tried to offset this community reaction. It immediately presented itself to the community to help residents and environmental groups understand and trust Getty's proposal. Getty's attitude could not have differed more from NCRA's. They were open, accessible, and acknowledged the environmental concerns at stake. They had proven themselves to be trustworthy and responsible in other interactions with the Bridger-Teton National Forest and environmental groups in another drilling project in nearby Fall Creek. Moreover, Getty had an "excellent" prospect, much

different from NCRA's questionable one. They were willing to compromise in order to drill (interview with Dick Hamilton, District Manager,
Getty Oil Company, November 1981).

Getty assured the town that its Little Granite proposal would not be
environmentally disastrous as the Sierra Club contended and would be a
financial bonus for the area. It reported that seismic data and geophysical analyses indicated a 44,900-acre anticline that could potentially hold
$1 billion worth of oil and gas: 50 million barrels of oil or 300 billion cubic
feet of natural gas. The State would receive $375 million if Getty's projec-
tions proved accurate. Teton County would receive one-third of whatever revenues were generated. (*Jackson Hole News*, October 15, 1981).
Getty assured the town that it would take whatever steps necessary to
protect the large elk herds as well as the other wildlife that Little Granite
Canyon supports as calving and grazing grounds and migration routes.
They minimized the probability of encountering hydrogen sulfide
("sour gas") and promised minimal impact on the county's or town's
cultural, social, or educational services. They assured that no drill rigs
would ever be visible from Jackson or the highways into town and that
operations would be quieter than usual because Getty would use
muffled generators and diesel-electric engines (*Jackson Hole Guide*, October 15, 1981).Getty's representative emphasized the company's record:
"Getty has demonstrated, in Teton County and elsewhere, that drilling
in sensitive environments can be accomplished successful" (*Jackson Hole
News*, October 15, 1981, pp. 3, 23).

Getty hoped that its strong public relations effort would encourage
a constructive dialogue with the environmentalists, the Jackson Hole
community, and the USFS/GS team. They hoped that they would be
able to address most concerns and still be able to explore and perhaps
develop their prospect (Hamilton interview, November, 1981). This dialogue never evolved. There seemed to be little incentive for the Sierra
Club or Jackson Hole community to negotiate; a Congressional solution
to the whole problem seemed imminent.

The Sierra Club was not willing to concede to drilling in a proposed
wilderness area, because the drilling would be precedent setting and
make it difficult to prevent further oil and gas exploration in other wilderness areas. There seemed little reason for them to negotiate with
Getty to mitigate impacts if, in fact, a no-development option still existed. Although congressional action was not guaranteed, it seemed likely.
In November, 1981, the Jackson Hole Alliance Director, Story Clark, felt
confident that "the gears for a solution are already in the works" (personal communication, November, 1981).

In late February, 1982, this optimism was shattered when the Wyoming Congressional delegation introduced its wilderness legislation to Congress. The Wyoming Wilderness Bill explicitly removed all designated wilderness areas in Wyoming from any oil and gas or other mining activities. It firmly set the boundaries of each wilderness area. To the shock of the Sierra Club, The Wilderness Society, the Wyoming Wilderness Association, Jackson Hole Alliance and many Jackson residents, the bill removed Little Granite Creek from the Gros Ventre Wilderness, theoretically paving the way for Getty's well.

The Congressional delegation's bill hardly put the issue to rest, though. By that time the dispute had been allowed to build to the point that positions had become entrenched. Therefore, the Jackson Hole Alliance, Sierra Club, and other environmental organizations, rather than responding favorably to Getty's advances, began looking for other means by which to oppose the well. And, as they found, the process is vulnerable; there are many avenues by which individuals and groups can influence decisions in the making and oppose decisions that do not accommodate their concerns.

III. THE PROCESS IS NOT DECISIVE

In 1962, the Outdoor Recreation Resources Review Commission commented on the Forest Service's ability to prevail in the midst of inevitable conflict over its decisions:

> The Forest Service does not stand alone in the face of pressures from one direction. One Chief of the Forest Service is alleged to have said, "I am supported by the pressures which surround me." With skillful manipulation, the various clientele groups tend to cancel out each others' efforts. To the extent that this occurs, the administrator is given greater discretion to make decisions which he considers to be in the public interest. (Robinson, 1975, p. 22)

Twenty-five years later, this description is no longer accurate. The Forest Service might have discretion to *make* decisions deemed to be in "the public interest," but these decisions are not supported by the conflicting pressures acting upon the agency. In 1987, decisions not deemed appropriate can be effectively opposed; they are not decisive.

Theoretically, a decision should be the final word on a matter; as Webster defines it, a *decision* is a "conclusion" or "a report of a conclusion." But, although Forest Service officials would certainly prefer that their decisions fit this definition, they frequently do not. Because the stakes are so great, groups whose concerns have not been accommodated by a decision inevitably oppose that decision. And, because the decision-making process is unable to conclusively determine which outcome is the appropriate one, these groups have grounds on which to make strong arguments against a decision. If an oil and gas leasing or permitting decision is not influenced in the making, there remain many different ways to potentially undo that decision once made.

Consider, for example, a case involving Montana's Bob Marshall Wilderness Area. After 4 years and four decisions, the fate of the permit and lease applications filed for the Bob Marshall were still up in the air. The first decision was made when the Region I Regional Forester decided to deny a prospecting permit for the area. This decision, though, was appealed by the permit applicant to the Forest Service Chief. The Chief disagreed with the Regional Forester's assessment and sent the application back to him to be reconsidered (*High Country News*, April 13, 1981). A second decision was then made when the Regional Forester reconsidered his original decision and again ruled against the applicant. But, the applicant again appealed. In the meantime, environmentalists concerned about the Bob Marshall's wilderness characteristics turned to Congress. A third decision was made when the House Committee on Interior and Insular Affairs evoked an emergency provision of the Federal Land Policy and Management Act of 1976 to withdraw the wilderness area from the mineral leasing laws (*High Country News*, May 29, 1981).

Once again, the dispute was not resolved. The Mountain States Legal Foundation and the Pacific Legal Foundation sued the Congressional committee, alleging that the committee's action was unconstitutional (529 F.Supp. 982). The fourth decision was made when the federal judge in this case ruled that the committee's action was constitutional, but only so long as the Secretary of the Interior set the time limit of the withdrawal (*New York Times*, December 17, 1981). The judge thereby forwarded responsibility for the "final" permit decision to the secretary. After four decisions and three decision-makers, the fate of oil and gas exploration in the Bob Marshall Wilderness was not yet "decided."

No decision is immune from opposition. Informal as well as formal policies are opposed. If a group is unsuccessful at influencing policies in the making, then they will oppose the policies when implemented. Opposition traditionally follows existing administrative and judicial channels. But opponents are hardly limited to these avenues; there is much room for an individual or group's creative instincts. If success is not achieved in the administrative process, then opponents may turn to the courts, Congress, the state, or other governing bodies; guerrilla tactics cannot be ruled out when conflict is permitted to develop to extremes. If differences are not resolved when prospecting decisions are being made, then the conflict flows over to leasing decisions, and then on to permitting decisions, and eventually to licensing decisions. In the end, however, just as in the Bob Marshall case, no mechanism is available to resolve disputes; no process exists to accommodate the interests at stake. Consistent with the long-held land-management paradigm, all the steps in the process are designed to inform the professional land manager's decision. Disputes fester and escalate as the parties jockey for position in subsequent rounds of appeal.

THE CONSERVATION OF CONFLICT

Reviewing the Palisades leasing case is a tedious task. Each stage is redundant; the issues argued are the same. But because the differences between several environmental organizations and the Forest Service were not addressed—the environmental interests were not accommodated to their satisfaction—the disputes persisted. The case proceeded through a succession of decisions that in the end merely served as transfer points, as the dispute moved from one forum to another. The process proceeds as if governed by a natural law of "conservation of conflict": the level of conflict either remains stable or increases as decisions move from one phase to the next. Seldom are attempts made to resolve conflicts and therefore defy this "natural" law.

Policymaking: The FPA Stipulation and Guidelines Dispute

Because of the intensifying interest in the oil and gas potential of the Western Overthrust Belt, the Forest Service concluded in RARE II that, even though the Palisades area had high-wilderness value, no decision could be made about its final status until more information was obtained about its oil and gas resources. Thus, the Palisades area was classified as neither wilderness nor nonwilderness, but instead placed into a third *further planning* category. In the Final Environmental Statement on its RARE-II wilderness evaluation process the U.S. Forest Service acknowledged its dilemma:

> Unless there is additional exploration for oil and gas resources permitted in many areas allocated to further planning, subsequent wilderness–nonwilderness decisions will have to rely on data not much better than currently exists.
>
> Exploration by drilling to determine oil and gas potential is essential in reaching conclusions in land management or project plans that allocate roadless areas.
>
> For the above reasons, oil and gas exploration (including drilling where adequate exploration requires it) will be considered an integral part of the

further planning process. (U.S. Department of Agriculture, Forest Service, 1979, p. 97)

In justifying its allocation decision in this way, the USFS was assuming that stipulations would protect surface resources against environmental impacts if exploration or development should be proposed. Additionally, they assumed that the Forest Service retains authority at later decision points (i.e., permitting and licensing) to control whatever activities may be proposed. But environmental groups questioned both assumptions and, therefore, the decisions reached.

The first task the Forest Service Washington Office officials undertook was developing a special stipulation to be attached to all leases issued in Further Planning Areas. The national environmental organizations closely followed the development of the FPA lease stipulation and FPA leasing and management guidelines. They wanted to ensure that the stipulation and guidelines were sufficient to protect an area's wilderness character should oil and gas exploration and development ever be proposed. Although the Washington Office distributed drafts of the agency's proposed stipulation and guidelines for comment before finalizing them, these groups felt that their involvement was merely *pro forma*. They did not believe that the U.S. Forest Service Washington Office ever took their criticisms and recommendations seriously (Interview with Phil Hocker, Sierra Club Board of Directors, November 1981, and Bruce Hamilton, Sierra Club Northern Great Plains Regional Representative, October 1981).

Sierra Club representatives criticized the proposed FPA stipulation on several grounds, going as far as drafting their own version that they felt would better assure maintenance of wilderness values while final allocation decisions were being made. They complained about the lack of public comment or discussion of the stipulation before it was promulgated, and expressed frustration at their inability to follow and participate in the process. They cited several unanswered letters to the USFS Washington Office on the matter. They urged the Forest Service "to either adopt the Sierra Club proposed revision, or at least to initiate a consultation procedure leading to major changes in the oil/gas administration of Further Planning Areas" (September 22, 1980 letter from Bruce Hamilton to USFS Chief R. Max Peterson). More specifically, Bruce Hamilton criticized the requirement of only an environmental assessment (EA), and not an EIS, stating that it was a contradiction of explicit RARE-II FES intention and "would completely void the integrity of the Further Planning Process" (Hamilton letter). But the U.S. Forest Service made no changes in the FPA Stipulation in response to these criticisms. Agency officials were confident that the FPA stipulation was adequate and enforceable as originally designed (interview with Syd Grey, USFS Minerals Specialist, Washington, D.C., August 1981). By failing to respond to the environmental groups' contentions, however, the agency ensured that the dispute would reappear when it came time to implement the stipulations.

The FPA Stipulation was supplemented with new *Forest Service Manual* (FSM) guidelines for managing the oil and gas resources in FPAs. These guidelines were also developed in the USFS Washington Office in consultation with those Forest Service Regions facing oil and gas pressures. One meeting was held in Washington in April 1980 to obtain input from environmentalists and energy-industry representatives. With respect to specific management of further planning areas, the guidelines emphasized that

> a primary reason for allocating an area to further planning was the need to gather additional data on which to base a wilderness–nonwilderness decision. Therefore, mineral exploration is considered an integral part of the further planning process but it must be conducted in such a way that a wilderness option is retained or can be restored by reclamation. (*Forest Service Manual*, Section 2822.14b)

The Forest Service's confidence in the ability of protective stipulations to guard against surface resource impacts was clear:

> Controls available in regulations and lease terms are generally sufficient to avoid environmental problems and protect wilderness values. (*Forest Service Manual* Section 2827.14b, Interim Directive No.6, December 31, 1980, p. 4)

In addition, the Forest Service felt assured that there remained other decision points where they could control lease activities and perhaps rectify leasing errors:

> While the prelease environmental analysis treats general issues and concerns (such as preservation of the wilderness option) that would seriously and necessarily be affected by lease operations, the operations stage is the time to address most concerns. (*Forest Service Manual*, Section 2822.14b)

But the national environmental organizations monitoring the development of these guidelines questioned the U.S. Forest Service's assumptions that lease stipulations would adequately protect surface resource values and that the USFS possessed sufficient authority at later decision points to control activities that proved threatening to wilderness characteristics.

In early September, 1980, USFS Chief R. Max Peterson distributed a draft of the FPA guidelines for comment from those participating in the April meeting. Bruce Hamilton, Sierra Club Northern Great Plains Regional Representative, immediately responded to the draft guidelines, expressing concern that the issues raised earlier by the Sierra Club and other environmental groups had not been "adequately addressed." He stressed that:

> We remain convinced (1) that the FPA stipulation is not adequate to preserve the wilderness option; (2) that highly environmentally sensitive zones in FPAs that can't be directionally drilled should not be leased; (3) that leasing prejudices the land allocation decision; and (4) that the Forest Service does not have a workable plan for making a timely unbiased land allocation decision in FPAs that are leased. (September 22, 1980 letter to USFS Chief Peterson)

Hamilton questioned the logic behind further leasing of national forest lands to determine their energy-resource potential, especially when a considerable amount of land was already leased.

On December 31, 1980, a policy "decision" was made when USFS Chief R. Max Peterson finalized the Further Planning Area guidelines. But the debate about the proper management of FPAs, especially with respect to oil and gas activities, was not put to rest. The environmental groups continued to question both the protection contained in the stipulations and the ability of the Forest Service to legally enforce these stipulations. They saw no reason to accept the guidelines as a *fait accompli.* They pursued their concerns in the Palisades leasing decision, again raising the same issues.

Policy Implementation: The Palisades Leasing Decision

When lease applications were filled for the Palisades Further Planning Area, the Region-II Forester decided to defer any decision until after the area's wilderness evaluations were completed and its status decided. But deferring the leasing decisions in this manner was not a costless option. With the Wilderness Act's leasing deadline rapidly approaching, inaction on the outstanding leases could have potentially severe repercussions for the oil and gas industry. The Mountain States Legal Foundation (MSLF), a nonprofit industry interest group, sued the Forest Service to force it to make a decision (499 F.Supp. 383 (D.Wyo. 1980)).

MSLF filed suit in Wyoming District Court arguing that the USFS inaction on the lease applications constituted:

> (1) a withdrawal of the lands from the operation of the Mineral Leasing Act of 1920, without submitting such withdrawal to Congress for approval as required by the Federal Land Policy and Management Act; and

> (2) a rule or regulation of either or both of the Departments of Interior and

> Agriculture which was not promulgated as required by the Administrative Procedure Act. (499 F.Supp. 385 (1980))

MSLF claimed that its members as well as the general public would be "irreparably injured by the delay or prevention of development of energy resources" (499 F.Supp. 386 (1980)) in the Palisades FPA. MSLF charged that the USFS' inaction would have "serious secondary impacts, such as increased unemployment, possible energy shortages and an increasing balance of trade deficit, which . . . affects the public's individual rights including the right of economic choice" (Brimmer, 1980, p. 1).

USFS and Department of the Interior (DOI) attorneys defended themselves in court by claiming that, although they had "proceeded slowly" on these leases, they had not specifically withdrawn the lands in question. Rather, they stressed that the Forest Service had been following other statutory mandates, specifically that of the Wilderness Act of 1964. The agency representatives expressed confidence that by deferring leasing decisions, they would be better able to make the appropriate decision at a later time when more information about the area's resources was available. They argued that there were simply too many unknowns at that time to confidently make a decision (Brimmer, 1980, p. 2).

But, on October 10, 1980, Wyoming District Court Judge Clarence Brimmer ruled:

> We cannot allow the Defendants to accomplish by inaction what they could not do by formal administrative order. By our decision herein, we do not purport to require the Secretary of the Interior to accept, reject, or even take action on the outstanding oil and gas leases. We merely hold that the action taken by the Secretary of Agriculture, in failing to act on the outstanding lease applications falls within the definition of withdrawal under 43 U.S.C. Section 1702(j) and the Secretary of Interior is required to notify Congress of such withdrawal or institute action on the applications. (Brimmer, 1980, p. 27)

With this ruling, the Forest Service was forced to make the leasing decisions, regardless of the potential impact on the area's wilderness characteristics.

The MSLF case fueled the dispute over the proper management of potential wilderness areas but did not resolve it. It moved the Forest Service "out of the frying pan and into the fire." It provided the agency with no guidance on how to decide, given the competing claims of the Sierra Club and the Mountain States Legal Foundation. But, it forced them to make a decision.

Because the Palisades area was one of the first FPAs with impending leasing decisions following Judge Brimmer's ruling, any decision reached in the case would be precedent setting. Now both national environmental organizations and the oil and gas industry associations were watching the USFS response closely. The Sierra Club, The Wilderness Society, the Wyoming Wilderness Association, and the Idaho Environmental Council closely monitored all USFS activities affecting the Palisades FPA. They had participated in the agency's wilderness evaluations from the beginning and had long advocated wilderness designation for the Palisades. They had been pleased when RARE I concluded in a wilderness recommendation for the Palisades; they were outraged when RARE II resulted in a further planning status because of the area's oil and gas potential. Their feeling was that the area's exceptional wilderness qualities stood for themselves; the area should be designated wilderness regardless of what energy resources, if any, were located beneath it.

In June, 1981, the Regional Forester recommended to the BLM that the Palisades leases be issued. On July 7, 1981, the Sierra Club filed an appeal of the Regional Forester's decision to recommend oil and gas leasing in the Palisades. The appeal immediately went before the USFS Chief. In their 50-page "Statement of Reasons in Support of Appeal" the Sierra Club outlined and defended their (by this time all-too-familiar) contention that:

> Leasing in the Palisades, as permitted by the Regional Forester's decision, will not preserve the wilderness option for the area. The decision commits the lands to non-wilderness uses, and will result in damage to the wilderness qualities of the Further Planning Area. The decision does not meet the criteria for environmental protection established by the RARE program. Nor does the decision assure that the Forest Service will obtain data on oil and gas resources which RARE II indicates is necessary to carry out land management planning. (Sheldon, 1980a, pp. 1–2)

The Sierra Club argued that the Forest Service decision did not satisfy the agency's own objectives in decision-making; it neither contributed to the agency's oil and gas resource information base with which to make a wilderness–nonwilderness decision nor preserved the wilderness option. Consequently, the environmental organization sought to prove that not only was the decision-making process flawed because it did not comply with NEPA's provisions, but also that the decision itself clearly did not satisfy the USFS' own objectives of resource mapping in FPAs while maintaining wilderness values. The Sierra Club feared that the Regional Forester's decision was a "heads they win, tails we lose" proposition: Regardless of the actual oil and gas potential of the Palisades, the wilderness option would likely be lost (Sheldon, 1980a).

The Regional Forester responded to the Sierra Club's *Statement-of-Reasons* point-by-point in supporting his decision. He viewed the Forest Service decision to be clearly justified by the *Forest Service Manual* guidelines. Further, he emphasized that his decision did not commit the Palisades to nonwilderness uses, but rather that it allowed the Forest Service, while protecting wilderness values with clear stipulations, to obtain information about oil and gas resources in the Palisades. The Regional Forester acknowledged that there were many uncertainties involved in making these decisions, but expressed confidence that later decision points involving specific development proposals would provide better information and involve less uncertainty (Sirmon, 1980).

The Sierra Club, in turn, responded point-by-point to the Regional Forester's response to their original statement, again disagreeing with the USFS' assertions. They continued to question whether or not the Forest Service actually possessed authority at later decision points to control oil and gas exploration and development activities as the Regional Forester was asserting. It was an administrative capability the Forest Service believed it possessed, but one that they had never exercised. The Sierra Club questioned the Regional Forester's assertion that the leasing decision was not a major federal action requiring an EIS. They argued that waiting for an application for a permit to drill (APD) before doing an EIS was an "incremental" approach to understanding and dealing with the issues and that "one of the primary purposes of NEPA was the elimination of just such bits and pieces decision-making. NEPA [required] a review of a proposal as a whole, before commitments of resources [were] made to it" (Sheldon, 1980b, p. 3).

After considering the arguments made by the Sierra Club and other affected parties in appeals documents as well as at an oral presentation held in Washington, D.C., USFS Chief Peterson reaffirmed the Regional Forester's decision on December 31, 1980, and forwarded the Forest Service recommendation on to the BLM (Johnson, 1981).

The Sierra Club continued to appeal, making the same arguments, through the BLM decision-making hierarchy. Once again, though, their efforts were in vain. Secretary of the Interior James Watt personally intercepted the appeal before it went before the Interior Board of Land Appeals. On May 28, 1981, after considering the Sierra Club's complaint and the Forest Service response, Watt decided that the Forest Service's recommendation was appropriate and issued the leases.

The debate was not put to rest. Although final authority to make decisions under the mineral leasing laws does rest with the Secretary of the Interior, the Secretary's decisions are just as susceptible to appeal as those of subordinates. Two days after Secretary Watt made his decision to issue the Palisades leases, the Sierra Club filed suit in Washington D.C. U.S. District Court against R. Max Peterson (USFS Chief), John R. Block (Secretary of Agriculture), Robert Burford (BLM Director) and James Watt. Not satisfied that their concerns had been addressed by the Forest Service and BLM/DOI appeals processes, and still convinced that their questions deserved attention, the Sierra Club alleged the failure of the various USFS, BLM, and DOI officials to fulfill their obligations under the National Environmental Policy Act.

In the conclusion to their "Motion for a Preliminary Injunction," Sierra Club attorneys highlighted their long-expressed and well-known concerns. Once again, the Sierra Club questioned the assumptions supporting the Forest Service's decision. Once again, it doubted that the Forest Service had the administrative capability to significantly control oil and gas activities once leases were issued. They saw no link between the RARE-II FES objective of obtaining further information about an area's oil and gas potential and this Forest Service decision. Finally, they once again called for a full EIS, hoping that more extensive analysis would make the consequences of the leasing decision more apparent to the Forest Service.

With the Sierra Club's lawsuit, the issues were moved to a new arena, once again to be debated and judged. Now the decision-maker, instead of being the USFS Regional Forester or Chief or the Secretary of the Interior, was a federal district court judge. The arguments presented by each group were the same; only the person listening was different.

On March 31, 1982, Washington, D.C., U.S. District Court Judge Aubrey E. Robinson ruled that the USFS did possess the authority to enforce lease stipulations, even when these stipulations may make exploration or development impossible:

> The lessees may legally obligate themselves to lease conditions that may result in the inability to explore or develop; that is knowing risk the lessees wish to take. (*Jackson Hole Guide*, April 15, 1982, p. 1)

More than 16 years after having been filed, the Palisades leases were issued. The final "decision" in the case was made by a federal district court judge. Judge Robinson's ruling gave a little to each party. The Forest Service was not required to do an EIS because its lease stipulations could effectively negate any development on the leaseholds. When an APD was filed for a leasehold 4 years later in the Mosquito Creek section of the Palisades area, environmental groups questioned the actual enforcement of these stipulations. The judge's ruling left the door open for this outcome. Although the leasing dispute seems resolved, the underlying issues about exploration in national forest roadless areas remain.

WHERE THERE'S A WILL THERE'S A WAY

The national forest management decision-making process is extremely vulnerable to delay and attack. The Palisades case followed the customary administrative appeals process, ending with judicial review. Not all cases are so neat and predictable, though. Whereas administrative avenues are seldom ignored, groups pursue other paths of least resistance where their influence is greatest. Consider the Union Pass Road upgrade case mentioned earlier. Once the Regional Forester announced that the Bridger-Teton National Forest plan's preferred alternative would provide for no upgrading of the Union Pass Road, Louisiana Pacific (L-P) took a different tack. The dispute entered another forum when Bob Baker, also a representative of Dubois to the Wyoming State Legislature, introduced a resolution opposing any reduction of timber harvesting in the Greater Yellowstone Ecosystem or any change of management objectives there. (The resolution passed the House but failed in the Senate.) He also tried to generate state pressure to upgrade the Union Pass road system as one part of a larger revision of transportation routes in Northwestern Wyoming. He held meetings with Paul Cleary, head of the Governor's Planning Office, who in turn announced:

> We'll review the Union Pass decision under the Forest Plan when the Plan is released to see to what extent people and goods would have to be transported into various parts of the Forest. Any improvements we might seek would be predicated on the demands of a transportation network in that area. (*Jackson Hole Guide*, March 6, 1986, p. A3)

Fred Kingwill immediately responded that:

> When the state does express itself, we pay attention to that, but we are federal land managers and we make the final decisions. If the state proposes that the road be upgraded to a state highway, we'll pay attention to that, but that doesn't mean that's the direction we'll take. We're not giving up our responsibilities to the state. (*Jackson Hole Guide*, March 6, 1986, p. A3)

On October 24, 1985, the Regional Forester decided to allow L-P to upgrade the road in order to transport lumber from the 2.5 mbf Little Sheep Mountain Sale to the Louisiana Pacific mill in Dubois. Upon completion of harvesting, however, they were to make the road impassable once again.

"We're back at war again" John Barlow lamented when Regional Forester Stan Tixier's decision to permit the road upgrade was announced (*Jackson Hole Guide*, October 29, 1985). The decision was immediately appealed by environmental interests to Forest Service Chief Max Peterson. After 8 months of deliberation and 5 hours of testimony in Washington, D.C., from representatives of all sides of the issue, the

Chief announced his decision. He not only allowed the road upgrading, but also left the final fate of the road to the Forest Service in the Forest Plan. Louisiana Pacific would not have to restore the road as the Regional Forester had ordered.

The conservationists were taken aback by Peterson's decision. Peterson's decision was made despite the fact that the yet-to-be-released BTNF plan's preferred alternative had the disputed road kept closed to logging trucks. In so doing, the decision only widened the communication chasm between the agency and conservationists and deepened their distrust of the agency.

The Chief's decision hardly ended the dispute, however. While the environmental groups frantically tried to obtain a temporary restraining order on road construction in order to pursue litigation, L-P began work on the road. By 9 a.m., when assistant Forest Supervisor Ernie Nunn arrived on the scene, a 27-foot-wide road had been cleared where only a 14-foot-wide section was allowed in the plans. Conservation groups' request for a restraining order had just been denied, the judge ruling that the impacts to wildlife and wildlife habitat were purely "speculative" (*High Country News*, July 7, 1986).

However, at the same time another temporary restraining order took affect, this one obtained by Joe and Stella Retel who own a summer home and property that is crossed by 300 feet of the Union Pass road. Whether or not the Forest Service has a legal right-of-way across the land was never established and Louisiana Pacific never negotiated one with the Retels. An angry Joe Retel said that even a half-million-dollar settlement offer by Louisiana Pacific would be turned down and commented that he was forcing the issue "for myself, the cattlemen and the independent loggers" (*High Country News*, July 7, 1986, p. 7).

Retel was aggravated that Louisiana Pacific just assumed they had a right-of-way across the Retel property: "The cowboys do us the courtesy of asking permission before they cross. But those lumber trucks aren't going to stop for you, a cow, for nobody" (*High Country News*, August 4, 1986, p. 3). The Retels are also pursuing the issue because of their concern for Upper Green wildlife. "Fish Creek and Teepee Creek used to be good fishing, but now all there is is silt," Joe Retel lamented. "If we could just stop Louisiana Pacific from timbering and tell 'em to get the hell out, that would be great" (*High Country News*, August 4, 1986, p. 3).

The Retel's attorney, John Zebre, underlined the concern of many at this point in the dispute that the Forest Service was listening only to Louisiana Pacific:

> The Forest Service now has the opportunity to demonstrate to the public that their sole interest in the Union Pass road is not the well-being of Louisiana Pacific. (*High Country News*, August 4, 1986, p. 3)

He wants them to "simply tell Louisiana Pacific that since Louisiana Pacific failed to secure the necessary right-of-way from a private landowner they should travel the route upon which they based their bid for the timber sale" (p. 3). Meanwhile, Forest Service officials were saying that "temporary log-hauling is a matter between Louisiana Pacific and the Retels" (p. 3).

Louisiana Pacific tested the Retel's resolve by sending its head forester, Bob Baker, and two logging trucks across the Retel's property. When the trucks returned laden with logs, Retel recalled, "I said to him (Baker), 'You're not going through.' He told me, 'I'm a public servant and I've got a right to go through'." Retel allowed Baker to pass "but the logs stay here" (*High Country News*, August 18, 1986, p. 4).

On August 18th, an agreement was signed by the Forest Service and the Retels, allowing Louisiana Pacific to haul the timber from the Little Sheep Mountain sale over the Retel's land. However, the agreement cuts short any consideration of requests "to haul additional timber sales across the Retel property until resolution of the permanent right-of-way and implementation of the forest plan." This restrictive wording was chosen, according to the Retel's attorney, because

> Louisiana Pacific has no intention of abiding by Mr. Peterson's decision, but rather after the Little Sheep Mountain sale, would seek additional timber harvest in the Upper Green and additional utilization of the road, and continue to make 'foot-in-the-door' arguments. (*High Country News*, September 1, 1986, p. 5)

Joe Retel summed up the agreement: "If they want to bid on another sale, they can go another way" (p. 5).

Getty Oil Company's proposed Little Granite Creek well provides another example of the resolve of interest groups who feel that the traditional administrative decision-making process has failed to address their concerns. Getty and a seismic testing firm in the area incurred more than $56,000 in damage at its construction site where surveying stakes were uprooted and expensive surveying equipment thrown into the creek. (*Jackson Hole Guide*, July 15, 1982) On July 4, Earth First! held a rally at the Little Granite Creek trailhead, protesting Getty's proposed well. 400 people attended the rally and 100 of those pledged to blockade Getty's access road should the oil company begin road construction. Author Edward Abbey (whose book *The Monkey Wrench Gang* inspired the founding of Earth First!) implored the gathering to "oppose, resist and, if necessary, subvert. . . . Earth first, grizzly bears second, people third, and J. Paul Getty last!" In concluding, Abbey joked with the audience to "please stop sending me those damned, dirty survey stakes!" (*Jackson Hole Guide*, July 8, 1982, p. A19).

CONCLUSION

Unfortunately, the cases discussed in this chapter by no means stand alone. Numerous others are now beginning or are in process. National forest management disputes are not generated by capricious Forest Service officials. The problem is *not* that the Forest Service selects the *wrong* alternatives to study, evaluates them using the *wrong* criteria, or assigns the *wrong* values to different outcomes. The problem is that there are often no *right* answers. Because these decisions are inherently subjective, numerous outcomes are both possible and legitimate. Likewise, any decision can be subject to question and opposition. Because the process is structured to develop technically defensible decisions when such decisions do not exist, it ensures that decisions will be opposed. The stakes are just too great to expect otherwise.

The next chapter explores some of the factors that contribute to public-land management problems as they exist today. Chapter 5 describes how the procedural and substantive requirements of the Forest Service have changed since the agency was established. Additionally, it addresses how the political environment of the 1960s and 1970s has changed the nature of public land management and how the Forest Service has responded to this change.

CHAPTER 5

Why the Paradigm Fails

Faced with protecting threatened Bighorn Sheep habitat yet accommodating energy exploration in Wyoming's Shoshone National Forest, Forest Supervisor Steve Mealey summarized his dilemma with these words:

> We face some tough decisions. The law asks bureaucrats to do things we don't know how to do sometimes. But that doesn't take us off the hook. (*High Country News*, November 29, 1983, p. 4)

Mealey, like all agency decision-makers, must balance the many competing interests with a stake in national forest management, and then render a decision. How he might successfully do so, however, has never been adequately prescribed.

As seen, national forest management in practice differs considerably from its statutory and administrative counterparts. In practice, the process is not sufficiently informative or convincing; it is divisive; and, moreover, it is not decisive. In short, the process fails in precisely those areas where, theoretically, it should excel. Why does the land management paradigm within which decision-making occurs now fail when it succeeded for so long?

Land management decisions are not purely technical. They cannot be solely subject to scientific review and analysis. On paper, however, the process hides the subjective aspects of decision-making under a cloak of technical expertise. It assumes that scientifically trained land managers will be able to acquire the appropriate information with which to analyze specific proposals and then reach outcomes that advance the public's interest. The inherent value judgments are hidden in technical analysis. Like the emperor's new clothes, however, this technical cloak now hides little and, in this instance at least, the masses are not quiet about what they see.

119

National forest management has never been an objective, purely technical task. Regardless, the professional land-management paradigm prevailed for the first half of the twentieth century; these decisions were long accepted to be scientific in nature and thus best left to professionals trained in scientific methods of management and decision-making. The paradigm remained intact because the Forest Service's professional judgment was widely accepted and, moreover, trusted. Compared with the alternative of unbridled commercial resource exploitation, the public welcomed professional forestry as a godsend.

Today, judgment exercised by professional foresters in managing public lands is no longer accepted; their management processes are no longer trusted. Several factors contribute to the changing environment of public land management. First, the Forest Service task is now more overtly subjective than it has ever been in the past. Congress has mandated numerous, often competing objectives that demand choices between legitimate yet conflicting outcomes. Furthermore, these mandates are not all consistent with the conservation ideal. Several are based in preservationist notions about appropriate natural resource management. Because the paradigm is premised on use, though, it does not easily accommodate preservation values. Additionally, the risks and uncertainty inherent in many aspects of national forest management have compounded the value judgments involved in decision-making.

The second factor changing the nature of public land management today is that the public has lost faith in the Forest Service's responsiveness to its concerns. Increasingly, groups and individuals are turning to Congress and the courts to obtain decisions that are deemed more appropriate and more just; decisions that accommodate their concerns. Finally, the numerous natural resource statutes enacted in the 1960s and 1970s have not only legitimized their arguments, but also distributed power among many user groups, who therefore can oppose decisions deemed inappropriate.

PUBLIC LAND MANAGEMENT IN TRANSITION

Today's U.S. Forest Service task differs markedly from that encountered when the agency was established at the turn of the twentieth century. The agency's paradigm, premised on scientific management, has always been subject to attack by some groups. After World War II, though, support began eroding at a much more rapid pace as other visions of appropriate public-land uses took hold and gained influence and power in decision-making. As the cases discussed in Chapter 4 illustrate, the Forest Service now confronts a political resource-allocation task in addition to its traditional scientific land-management responsibilities. Although the problem framed by the public and Congress has changed, the decision-making process applied by the Forest Service has remained grounded in the same paradigm; one wherein professionals are responsible for acquiring and assimilating information and pronouncing efficient outcomes; one wherein conservation, not preservation, of resources is the end. The same paradigm is unable to accommodate both ends to the satisfaction of the adherents of each.

The scenario described in Chapter 4 did not exist prior to World War II. Before that time, the U.S. Forest Service served largely as custodian of the national forests, using professional silvicultural and pest control, and flood and fire control practices to maintain the forests. At the time, there was little demand for national forest resources with the exception of timber, and that was still marginal. The agency concentrated on assisting private forest management as well as fire, pest, and disease control in public and private forests. Part of its responsibility was road-building to promote firefighting, overseeing the forests, and logging when it did occur. What disputes arose were the old ones between preservationists advocating wilderness preservation and the forestry conservationists advocating resource use (Frome, 1974a,b; Steen, 1976). During this time, preservationists had little influence in Forest Service decision-making or recourse against decisions they deemed inappropriate.

THE HEART OF THE PARADIGM CHALLENGED

Although preservationists had supported the early efforts of Pinchot and the forestry profession to establish and manage the public forests, they never adopted the conservation philosophy themselves. At the time, the conservationist's proposal was simply a more desireable alternative to the nonmanagement and disposal problems then plaguing the forests. Moreover, it had more political and popular support than did the preservationists' proposals. Regardless, a major difference in values continued to separate the two groups and create tensions between them. Their disagreements were highlighted by the battle over damming the Hetch Hetchy valley in California's Sierra Nevada range. This dispute destroyed the long friendship between John Muir and Gifford Pinchot. Preservationists continued to try to influence Pinchot and his successors to consider preserving some national forests intact for the benefit of future generations. The two philosophies were incompatible, though, and the preservationists' efforts met with little success.

For much of United States history, wilderness was something to be conquered and subdued. As historian Roderick Nash (1967) described this early view:

> The pioneers' situation and attitude prompted them to use military metaphors to discuss the coming of civilization. Countless diaries, addresses, and memorials of the frontier period represented wilderness as an "enemy" which had to be "conquered," "subdued," and "vanquished" by a "pioneer army." The same phraseology persisted into the present century; an old Michigan pioneer recalled how as a youth he had engaged in a "struggle with nature" for the purpose of "converting a wilderness into a rich and prosperous civilization." Historians of westward expansion chose the same figure: "they conquered the wilderness, they subdued the forests, they reduced the land to fruitful subjection." (p. 27)

Once the Forest Service was well-established, however, a minority of federal land managers, private individuals, and groups again began advocating the preservation for noncommercial purposes of several particularly scenic or ecologically important areas. Arthur Carhart, the Forest Service's first landscape architect, was assigned the task of planning for recreation and summer-home development in particularly scenic national forest areas. He reported back to his superiors that:

> the first logical step in any work of this type is to plan for preservation and protection of all of those things that are of values great enough to sacrifice a certain amount of economic return for esthetic qualities. (Frome, 1974a, p. 118)

This comment, coming in 1919, was one of the first intraagency indications that preservation might have definable values that should not be universally disregarded in favor of the quantifiable commercial land uses. It also conformed to Forest Service policies regarding economic efficiency in land management. It was not readily accepted, however.

Several years later, U.S. Forest Ranger Aldo Leopold pursued the issues Carhart had raised. Leopold's efforts led to administrative classification of the first national forest wilderness area, the Gila Wilderness in the Southwest (Frome, 1974a, pp. 118–120; Steen, 1976, pp. 154–155). Forest Service Chief William B. Greeley not only approved the Gila Wilderness designation but also encouraged further designations in other national forests. He commented that "the frontier has long ceased to be a barrier to civilization. The question is rather how much of it should be kept to preserve our civilization" (Frome, 1974a, p. 121). In 1929, Greeley established Regulation L-20 under which national forest *primitive areas* could be designated by the Chief. These primitive areas were intended to maintain "primitive conditions of environment, transportation, habitation, and subsistence, with a view to conserving the value of such areas for purposes of public education, inspiration and recreation." Exceptions to these purposes could be authorized by either the Chief or the Secretary of Agriculture. (Frome, 1974a, p. 121; Steen, 1976, pp. 154–156)

Preservationists were not appeased by this new Forest Service policy. The Forest Service values in preservation did not coincide with their own. In his review and analysis of these early regulations, historian Michael Frome questioned the true Forest Service intent:

> The Forest Service, however, did not anticipate reserving the primitive areas indefinitely from commercial use. Many of the remote portions with outstanding scenic and recreational qualities were being kept from haphazard road-building and commercial development until a time when more intensive study was needed. It is also conceivable that the Forest Service was trying to keep one step ahead of its "sister agency," the National Park Service; by demonstrating active concern for wilderness, it was better able to block establishment of new parks out of old forests. (Frome, 1974a, p. 121)

Perceiving a wolf in sheep's clothing, The Wilderness Society was established in 1934 "to secure the preservation of wilderness, conduct educational programs concerning the value of wilderness, encourage scientific studies, and mobilize cooperation in resisting the invasion of wilderness" (Frome, 1974a, p. 123). Organization and pressure external to the Forest Service was needed because national forest wilderness areas were being threatened more than at any time in the past. One of the organization's founders was Bob Marshall, then Director of Forestry for the Department of the Interior's Office of Indian Affairs. In that capacity, he initiated a series of memos between himself and then Secretary of the Interior Harold Ickes about the potential consequences of New Deal public works programs on undeveloped wilderness areas. Marshall's greatest worry was that roads would soon traverse most undeveloped territory and he recommended to Ickes that wilderness areas be set aside with specific standards prohibiting development within their boundaries (Frome, 1974a, p. 105).

In 1937, Marshall continued his battles for wilderness preservation as Forest Service Chief of Recreation and Lands. His efforts, combined with continuing pressure from constituencies such as The Wilderness Society, led to Forest Service "U" regulations in September 1939. These regulations were stronger than the old L-20 regulations and "established a procedure for expansion of wilderness and for excluding developments previously permissible in primitive areas" (Frome, 1974a, p. 125). Forest Service administrative classifications under these "U" regulations remained few and far between, though. Many proposals encountered opposition from USFS officials who perceived their responsibility to be one of managing the public lands for multiple-uses, not preservation of a single use. As a result, the remaining national forest wilderness areas were gradually being consumed by other uses. In 1926, Forest Service figures showed the national forest system contained 74 roadless areas, each having at least 230,400 acres. The largest roadless area at that time was 7 million acres in size. In total, the 74 areas covered 55 million acres. By 1961, similar reviews indicated that only 19 roadless areas of 230,400 acres or more remained. The largest roadless area at that time had dwindled to 2 million acres. All 19 areas now totaled only 17 million acres (Frome, 1974a, pp. 20–21).

Wilderness areas are distinguished by their roadless feature. However, roads are critical to timber and other commercial development as well as to fire prevention and control. As these pressures intensified, wilderness was destroyed and the national forest road system exploded. Consequently, those lands set aside as wilderness were most often those having little or no value for other uses. As one historian of the long fought battle over wilderness recounted:

> The protected wilderness existed more by accident than design. Most of its commercial resources, composed of lands by-passed in the rush of settlement and exploitation from east to west, were too poor to utilize or too costly to develop. About one-fourth of all acreage in reserved wilderness was composed of mountain peaks, desert, sand dunes, lava flows, and rock slides; about one-third was covered with brush or with scrubby and other nonproductive forests; another third was productive timberland, while a small percentage was meadow, grassland, or water surface. The timberlands contained 8 million acres of productive wood sources—only about 2 percent of the nation's total, or a sufficient volume to supply national needs for two years. (Frome, 1974a, p. 21)

As demands for national forest resources exploded following World War II, the Forest Service "U" regulations proved insufficient to protect wilderness areas to the satisfaction of preservationists. They perceived in the Forest Service an emerging emphasis on timber production at the expense of wilderness and other resource values. As a result, what wilderness had been protected began succumbing to commercial development pressures. Administratively designated areas were split by roads and some areas were logged. Critically, the Forest Service was also raising the ire of the general public, as well as preservationists, as its clear cutting practices increased. This increasing clamor helped bolster the preservationists' arguments and open Congress's door. Michael

Frome (1974a) reviewed the events that led to diminished confidence in administrative management of wilderness areas and in the Forest Service in general:

> Although citizen groups had long supported the Forest Service as an agency concerned with scenic resources and wilderness, they lost their place as a key part of its constituency. The timber-first policy came to the fore in response to several factors, one being the political pressure of the timber industry, which, having intensively cut its own private lands without adequate concern for sustained yield, became reliant on public lands, including the remaining virgin forests, to keep its mills going. (p. 132)

After World War II, the problem currently posed by public land management emerged. The resource allocation dimension of the Forest Service task came to the forefront of decisions as conflicting demands for the varied national forest resources increased. Timber demands were increased by an industry that had both poorly managed the private forests and needed expanded sources of timber to meet the burgeoning demands of a post-war housing boom. The amount of timber cut from the national forests increased from approximately 500 million board feet (1%–2% of total domestic production) in 1910, to about 2 billion board feet just before World War II, and more than 10 billion board feet by 1965 (Clawson, 1967, p. 8). Recreation demands exploded from a population with increased leisure time. There were approximately 5 million recreational visits to the national forest in 1925 compared to 150 million by 1965 (Clawson, 1967, p. 10). An increasingly urbanized society demanded more opportunities for solitude and wilderness retreat. The nation's economic productivity was increasing along with the demand for the mineral resources contained in the public lands. The environmental awakening of the 1960s led to demands for increased wilderness preservation, wildlife protection, and watershed management. It also led to an increased concern about humanity's responsibility toward and dependence upon the natural environment. And, in the 1970s, demands for domestic energy production increased and attention turned to the previously neglected national forests.

Groups and individuals concerned with the noncommercial scenic, wilderness, and wildlife resources were not satisfied with the Forest Service's response to their demands. They believed that too much emphasis was being placed on timber sales and road-building to accommodate logging, all at the expense of particularly scenic and wild areas. With their concerns unaccommodated by the administrative process, these groups shifted their attention to Congress.

THE CONGRESSIONAL RESPONSE TO CHANGING DEMANDS

Initially, Congress responded cautiously. The first problem raised was the age-old question of wilderness preservation. Brought before Congress in the mid-1950s, the Wilderness Act was not enacted until 1964. By then, the clamor was intensified and, as might have been predicted, Congress became more responsive. In fact, the environmental awakening and, moreover, the social activism of the late 1960s and early 1970s provided Congress with little choice. As one analyst of the environmental decade commented:

> Although Nixon was by no means an enthusiastic supporter of the environmental movement, his signature on NEPA was considered indicative of the fact that no politician could afford to ignore the demands being made by the movement. (Wenner, 1982, p. 1)

Similarly, Senator Edmund Muskie's adoption of the clean-air cause has been attributed by some political analysts as an attempt to respond to a no-lose growing public concern, thus gaining support for his presidential campaign (Yaffee, 1982, p. 151). The environment was a cause to advance; the decade saw the passage of several monumental environmental statutes that have destroyed the original premise upon which the traditional land-management paradigm is based: conservation.

The issues placed before Congress are hardly simple ones to resolve. They are not, as often portrayed, obvious questions of good (the public interest) against evil (industry and an unresponsive bureaucracy). Arguments for more wilderness preservation mean less timber and mineral development, both critical to a thriving economy. As is the nature of the legislative process, compromise is necessary. Because each constituency has its advocates in Congress, legislation can seldom be enacted before it contains something for each. The result was a collection of natural resource statutes that contain obviously conflicting mandates to public land managers. And, as seen in the cases in Chapter 4, because objectives compete, there is no technically correct decision; almost any decision reached can be opposed on the grounds that it fails to address the objectives of a particular statute.

Several natural resource and environmental statutes directly affect national forest management today. These statutes are outlined below, with special emphasis given to those provisions and requirements that cloud Forest Service decision-making and give the power to land-user groups to question administrative decisions deemed inappropriate. Particular emphasis is placed on the Wilderness Act of 1964, because Forest

Service implementation of this act's provisions fuel many disputes involving oil and gas exploration, road-building and timber-harvesting in the national forests and muddies the agency's long range planning efforts.

THE WILDERNESS ACT OF 1964

Because the early USFS wilderness classifications were administrative decisions, they could easily be undone. As a result, preservationists feared that future administrators might reclassify land in response to commercial development pressures. In the eyes of preservationists, the administrative classification system also suffered from one other important defect; because it was haphazard, it did not ensure that important wilderness areas would *ever* be preserved. Consequently, preservationists moved their efforts to the Congressional arena in the 1950s. They hoped to encourage legislation that would both institutionalize and systematize the preservation process, as well as preserve designated wilderness areas in perpetuity.

The Forest Service opposed the act, arguing that:

> This bill would give a degree of Congressional protection to wilderness use of the national forests which is not enjoyed by any other use. It would tend to hamper free and effective application of administrative judgment which now determines, and should continue to determine, the use or combination of uses to which a particular national forest is put. (Frome, 1974a, p. 140)

The Forest Service was joined by the National Park Service, the two professional forestry associations—the Society of American Foresters and the American Forestry Association—the forest products industry, the oil and mining industries, and grazing interests in opposing the act. Eighteen hearings were held on the Wilderness Bill between 1957, when it was first introduced, and 1964, when it finally passed. The bill was rewritten time and again before it was accepted by all affected groups (Frome, 1974a; McCloskey, 1966). In September 1964, President Johnson signed the Wilderness Act into law.

The Wilderness Act of 1964 (16 U.S.C. 1121(note)) largely accomplished the preservationists' objectives. In the act, Congress declared that:

> In order to assure that an increasing population . . . does not occupy and modify all areas within the United States . . . leaving no lands designated for preservation and protection in their natural condition, it is hereby declared to be the policy of Congress to secure for the American people . . . an enduring resource of wilderness. For this purpose it is hereby established a National

Wilderness Preservation System to be composed of federally owned areas designated by Congress as "wilderness area." (Section 2(a))

The act automatically placed all administratively classified national forest *Wilderness, Wild,* and *Canoe* areas in the new wilderness system. It instructed the Secretary of Agriculture or Chief of the U.S. Forest Service to review all administratively designated "primitive" areas for possible inclusion in the system. A 10-year deadline was imposed for reporting their findings to the President. The President was charged with making recommendations to Congress regarding which "primitive" lands should become Wilderness. Congress is the final decision-maker on all Wilderness designations. Similar instructions were spelled out for the Secretary of the Interior with respect to roadless areas in the national park system and national wildlife refuges. (In 1976, the Federal Land Policy and Management Act—commonly referred to as the BLM "Organic" Act—extended wilderness evaluation and designation mandates to BLM lands [43 U.S.C. 1701(note)].)

Congress provided a very lengthy definition of the term *wilderness* in order to guide the two Secretaries, the President, and eventually itself in making wilderness designation recommendations and decisions:

A wilderness in contrast with those areas where man and his own works dominate the landscape, is hereby recognized as an area where the earth and its community of life are untrammeled by man, where man himself is a visitor who does not remain . . . wilderness is further defined to mean . . . an area of undeveloped Federal land retaining its primeval character and influence, without permanent improvements or human habitation, which is protected and managed so as to preserve its natural conditions and which (1) generally appears to have been affected primarily by the forces of nature, with the imprint of man's work substantially unnoticeable; (2) has outstanding opportunities for solitude or a primitive and unconfined type of recreation; (3) has at least five thousand acres of land or is of sufficient size as to make practicable its preservation and use in an unimpaired condition; and, (4) may also contain ecological, geological, or other features of scientific, educational, scenic, or historical value. (Section 2(c))

The Secretaries of the Interior and Agriculture were instructed by Congress to manage Wilderness in order "to preserve its wilderness character . . . wilderness areas shall be devoted to the public purposes of recreational, scenic, scientific, educational, conservation and historical use" (Section 4(b)). And the act clearly prohibited certain uses: ". . . there shall be no commercial enterprise and no permanent road within any wilderness area . . . no temporary road, no use of motor vehicles, motorized equipment or motorboats, no landing of aircraft, no other form of mechanical transport, and no structure or installation within any such area" (Section 4(c)). These prohibitions were not absolute; exceptions were permitted where necessary to administer the act's provisions and manage the wilderness area, as well as in the case of emergencies which threatened the health or safety of individuals.

Whereas the Wilderness Act created a Wilderness Preservation System that is much more extensive than its proponents ever imagined (Bury and Lapotka, 1979, pp. 12–20), it also permitted activities within this system that run counter to the wilderness ideal. The Wilderness Act was the product of 9 years of Congressional debate, negotiation, and compromise. The major concession made by preservationists—and the concession that finally led to the act's passage—was Section 4(d)3. This section permitted mineral leasing in national forest wilderness areas until December 31, 1983. This provision is responsible for considerable conflict today about how and where oil and gas exploration and development may occur on public lands. Section 4(d)3 specifically permits:

> use of the land for mineral location and development and exploration, drilling, and production, and use of land for transmission lines, waterlines, telephone lines, or facilities necessary in exploring, drilling, producing, mining, and processing operations, including where essential the use of mechanized ground or air equipment and restoration as near as practicable of the surface of the land disturbed in performing prospecting, location, and, in oil and gas leasing, discovery work, exploration, drilling, and production, as soon as they have served their purpose.

The provision permits the Secretary of Agriculture to attach "reasonable stipulations" to mineral leases, permits, and licenses in Wilderness areas "for the protection of the wilderness character of the land" (Section 4(d)3).

Initially, this provision posed little threat to Wilderness; today, it is creating havoc. Several major oil and gas discoveries have attracted industry into areas previously thought to hold little energy potential, and improved technology has permitted exploration of previously inaccessible yet promising formations. As a result, several national forests, particularly those along the Western Overthrust Belt, have experienced tremendous pressure to permit energy exploration within both designated and proposed Wilderness Areas. In implementing the Wilderness Act's provisions, the Forest Service initially followed an unwritten policy of *not* issuing leases in Wilderness Areas. In the early 1980s, this policy was predictably challenged (U.S. General Accounting Office, 1982, p. 19).

Regardless of Section 4(d)3's legislative legitimacy, preservationists still oppose energy or mineral development in Wilderness Areas unless justified by national emergency. They argue that the protective phrases of the act are too weak. (Some wilderness areas contain leases issued before their designation as Wilderness. Consequently, these leases have no protective stipulations attached to them. The government lacks power to attach stipulations after the fact without agreement of the lessee. Otherwise, such an act would constitute a breach of contract on the part of the government. Lessees obviously lack incentives to agree to such conditions.) Preservationists also question the rationale behind the phrase "restoration as near as practicable" of mined or drilled lands. They argue that if the land cannot be restored to its wilderness state then it should not be intruded upon at all.

Finally, preservationists note that the December 31, 1983, deadline still permits exploration and production after that date. Leases issued on December 31, 1983, are valid for 10 years. During that period, the lessee has a legal right to develop the oil and gas resources, even though the leasehold is located in a Wilderness. Hypothetically, lessees could hold their leases for 9½ years before submitting an application for a permit to drill. If the permit is granted, the clock is stopped until exploration and production, if warranted, are completed. Consequently, oil and gas production in Wilderness areas could easily extend well beyond the year 2000, long after the act's December 31, 1983, deadline for issuing new leases. December 31, 1983, hardly marked the end to this debate.

The Forest Service's own timber-harvesting program has similarly put pressure on the wilderness-designation process. As the amount of timber available from private lands continues to decrease, industry interest in national forest supplies is increasing. Future timber harvests and the associated road building are targeted at those areas where good timber supplies still exist, including roadless areas. One recent environmental group newsletter warned of the threat to potential wilderness areas:

> In preparation for the anticipated flurry of timber sales, the Forest Service has been building roads. Already there are 342,000 miles of roads in the national forests (compared to 42,500 miles in the interstate highway system), and the service plans to build 58,000 miles of new roads over the next 15 years. Conservationists are particularly concerned about diminishing roadless areas. About 20 percent of the planned roads will be put in pristine areas. (*National Audubon Society News Journal*, June, 1986, p. 7)

Once wilderness preservation became law, the Forest Service was forced to deal with these lands on Congress's terms. Congress's message, however, was still vague and left much to the Forest Service's discretion. The Wilderness Act, couched in laudable but broad terms, left much undefined. The two decades following its passage have been filled with debate over whether or not specific areas fall under the broad provisions of the act. Debate has raged over when an area "generally appears" to be in its natural state with the work of man "substantially unnoticeable." These determinations are highly subjective; what is "substantially unnoticeable" to one evaluator can be a glaring defect to another. Debate has also centered on when an area contains "features of scientific, educational, scenic or historical value" (Udall, 1978, pp. 32–34). Again, making this determination requires judgment about the relative values of an area; when do commercial values outweigh scenic values? The Forest Service view has differed from that of preservationists:

> The agency committed itself to impeccable compliance with the letter of the law and fulfilled it thoroughly insofar as the primitive areas and wilderness areas are concerned. Its attitude toward potential additional areas, or even to consideration of any, is quite another story, one of consistent resistance to proposals of new wilderness, endless confrontations with citizen groups in virtually all parts of the country, often ending with officially sanctioned intrusion and commercial exploitation of the contested regions so as to render the wilderness "invalid." (Frome, 1974a, p. 153)

Disagreement over the act's intent led to two massive Roadless Area Review and Evaluation programs (RARE I and RARE II) by the USFS in which *all* national forest roadless areas, not just the administratively classified primitive areas, were evaluated for possible inclusion in the Wilderness system (Johnson, 1979; Merritt, 1975; Sumner, 1977). Because of a 1982 U.S. District Court ruling in California that the Forest Service RARE-II EIS is inadequate, the Forest Service announced in February, 1983, its plans to scrap RARE II and commence a new, RARE-III process (*High Country News*, February 4, 1983). This process would be undertaken only in those states that have no comprehensive wilderness legislation of their own. The announcement led to a flurry of wilderness bills introduced in Congress by state congressional delegations trying to avoid continued uncertainty over commercial resource use should their states be subjected to a third roadless evaluation. Debate over the fate of those roadless areas not formally designated Wilderness yet still retaining wilderness attributes held up much of this legislation for more than a year and still clouds agency decision-making, particularly in its current comprehensive forest-planning effort (*San Francisco Chronicle*, May 3, 1984). Timber sale proposals in many areas have been put on hold while the fate of both proposed wildernesses and released roadless lands is decided, (*Jackson Hole Guide*, February 9, 1984).

Whereas questions of wilderness designation and management have dominated many oil- and gas-leasing and timber-harvesting disputes, other statutes have an important role in Forest Service decision-making. Combined, these acts increase Forest Service discretion at the same time as increasing the influence over decision-making of different public land user groups.

THE WILD AND SCENIC RIVERS ACT

> The great rivers of this country represent vestiges of a frontier America where waterways were the highways to exploration and development. Today, these wondrous resources have fallen victim to excessive industrialization, abusive land use, and an overall move to commercialize the recreational value of free-flowing rivers. . . . [they] are now in danger of extinction. (Goodell, 1978, p. 43)

There are few remaining river systems in the United States that are unencumbered from their headwaters to their mouth. Responding to the unbridled development, particularly for power generation, of the nation's few remaining free-flowing rivers, Congress enacted the Wild and Scenic Rivers Act in October 1968 (16 U.S.C. 1271–1287). The battles leading to its enactment were not as lengthy as those for the Wilderness Act but they were fought by the same groups over the same ideological issues. Like the Wilderness Act, the Wild and Scenic Rivers Act established a "Wild and Scenic Rivers System." Unlike the National Wilderness Preservation System, though, there were three different components of the Wild and Scenic Rivers System–wild, scenic, and recreational rivers—each receiving different levels of protection. Once again, Congress expressed its laudable yet broad policy objectives to be implemented by federal public land management agencies:

> It is hereby declared to be the policy of the United States that certain selected rivers of the Nation which, with their immediate environments, possess outstandingly remarkable scenic, recreational, geologic, fish and wildlife, historic, cultural, or other similar values, shall be preserved in free-flowing condition, and that they and their immediate environments shall be protected for the benefit and enjoyment of present and future generations. (16 U.S.C. 1271)

In enacting this legislation, Congress was especially concerned about rivers where "there is the greatest likelihood of developments which, if undertaken, would render the rivers unsuitable for inclusion in the national wild and scenic rivers system" (16 U.S.C. 1275). Thus, the act required the designated federal agencies to make decisions in areas where conflict between different resource demands was most intense.

In administering the act, Congress mandated that all federal agencies give "primary emphasis" to protecting the "esthetic, scenic, historic, archaeologic, and scientific features" of these rivers. A protected river is to be administered "in such manner as to protect and enhance the values that caused it to be included in said system" (16 U.S.C. 1281). Congress defined *wild rivers* as:

> Those rivers or sections of rivers that are free of impoundments and generally inaccessible except by trail, with watersheds or shorelines essentially primitive and waters unpolluted. These represent the vestiges of primitive America. (16 U.S.C. 1273)

Scenic rivers were defined as:

> Those rivers or sections of rivers that are free of impoundments, with shorelines or watersheds still largely primitive and shorelines largely undeveloped, but accessible in places by roads. (16 U.S.C. 1273)

Finally, *recreational rivers* were defined as:

> Those rivers or sections of rivers that are readily accessible by road or railroad, that may have some development along their shorelines, and that may have undergone some impoundment or diversion in the past. (16 U.S.C. 1273)

The Wild and Scenic Rivers Act withdraws one-quarter of a mile of land on either side of a wild river from the mineral leasing laws. Scenic and recreational rivers remain under the provisions of the mineral leasing laws but subject to conditions imposed by the Secretaries of Interior and Agriculture to safeguard the areas. Potential wild rivers can still be leased but only with sufficient safeguards to protect them should they eventually become part of the Wild and Scenic Rivers System.

Because those areas where noncommercial resource values conflict with both timber harvesting and oil and gas exploration are the few remaining wild areas on national forest lands, it follows that these same lands, having not yet been exposed to commercial development activities, will have relatively pure rivers running through them. The Salmon River is a congressionally-designated Wild and Scenic River that was threatened by proposed construction of a timber road in Idaho's Nez Perce National Forest and led to a series of administrative appeals and lawsuits by environmental groups. Similarly, the Snake River is a proposed Wild River that caused environmental concern over leasing in Wyoming's Palisades area and drilling in Little Granite Creek. Consequently, proposals to develop the commercial resources of these areas inevitably are opposed by groups and individuals advocating a wild river's designation and protection.

Once again, the federal public-land managers are caught in the middle, forced to make decisions when the objectives they are to satisfy clearly conflict. Once again, broad standards such as "outstandingly remarkable" values are subject to debate between those favoring and those opposing a river's designation.

THE ENDANGERED SPECIES ACT

In 1973, Congress enacted the Endangered Species Act (16 U.S.C. 1531–1543). The provisions of this act have a role that surpasses that of the Wilderness Act or the Wild and Scenic Rivers Act in decision-making for public lands. This act provides for two categories of species: endangered and threatened. Section 7 of the act requires federal agencies to take actions necessary to protect endangered or threatened species habitat. Thus, the U.S. Forest Service must evaluate the effects of its timber harvesting and oil and gas leasing or exploration decisions on endangered or threatened species in what is called a "biological review." If their assessment concludes that a critical habitat may be modified or destroyed, the agencies are required to consult with the U.S. Fish and Wildlife Service (FWS). When consultation is required, the FWS's regional director completes a "threshold examination" to determine what effect, if any, the proposed activity will have on a listed species or its critical habitat. No further consultation between the USFS and FWS is required if this examination indicates little or no threat to the species. Should the proposed activity jeopardize a species, however, the FWS renders a "biological opinion" developed from information provided by the U.S. Forest Service. This opinion includes recommended modifications to the proposed activity to protect the species. Although the USFS is not bound to comply with the findings and recommendations in this opinion, failure to do so can lead to court action by either the Fish and Wildlife Service or a private group or individual (Yaffee, 1982, pp. 98–99). Section 11(g) provides for citizen lawsuits when an agency is not perceived to be fulfilling its responsibilities.

Current case law has interpreted Section 7 of the act to imply "an explicit Congressional decision to afford first priority to the declared national policy of saving endangered species" (437 U.S. 153, 185). Therefore, endangered species protection is to be paramount in land-management decision-making. Moreover, in one recent case the judge ruled that

> any contract which [the Secretary] enters into (e.g., a lease) which requires a future action on his part (e.g., approval of plans) will contain as an *implied term* a condition that the Secretary will behave lawfully (e.g., not violate the Endangered Species Act). (623 F.2d 712, 715; Martin, 1982)

In other words, under the Endangered Species Act an APD can conceivably be denied even though a lessee holds the rights to an area's oil and gas resources.

As timber-harvesting pressures and oil and gas industry interest spread to the more primitive and wild public lands, the Endangered Species Act plays a much more critical role in land-management decision-making. Environmental groups successfully contested the Forest Service failure to obtain a biological opinion from the Fish and Wildlife Service about the effect on the endangered Rocky Mountain Gray Wolf of proposed timber harvesting and road-building in the Jersey-Jack area of the Nez Perce National Forest. A lawsuit was similarly filed by environmental groups questioning the effects of Forest Service management of a Texas national forest on the endangered Red-Cockaded Woodpecker. Exploration proposals in Montana's Bob Marshall Wilderness have been opposed because they threaten one of few remaining habitats of the endangered Grizzly Bear (Findley, 1982, pp. 310–313). Leasing proposals in a Wyoming wilderness-study area potentially threaten the endangered or threatened Grizzly Bear, Bald Eagle, and Peregrine Falcon (U.S. Department of Agriculture, Forest Service, 1980). Leasing proposals in a California national forest threatened the only remaining California Condor nesting area.

THE FEDERAL CLEAN AIR AND WATER ACTS

Additionally, certain provisions of the Clean Air Act (42 U.S.C. 7409) and the Federal Water Pollution Control Act (as amended by the Clean Water Act) (33 U.S.C. 1324) affect Forest Service land-management decisions. Many proposed areas subject to mineral-leasing and timber-harvesting disputes are in those undeveloped, pristine backcountry areas that have been assigned "Class I" status under both acts. This status means that existing air- and water-quality levels cannot be "significantly" deteriorated or degraded (U.S. Department of Agriculture, 1981).

MULTIPLE-USE MANDATES AND PROCEDURAL REQUIREMENTS

In addition to considering the preservation mandates of each of the previously described statutes, the Forest Service must also fulfill the broader mandates of the Multiple-Use Sustained-Yield Act of 1960 (16 U.S.C. 528–531). And, to ensure that environmental concerns are considered in decision-making, Congress passed the National Environmental Policy Act in 1969 (42 U.S.C. 4321–4347) and National Forest Management Act in 1976 (16 U.S.C. 1600(note)). Both acts impose procedural requirements as a way of encouraging an understanding of and planning for the long-term consequences of particular decisions.

In the 1960 Multiple-Use Sustained-Yield Act, Congress established its policy that the national forests "shall be administered for outdoor recreation, range, timber, watershed, and wildlife and fish purposes" (16 U.S.C. 528). The act was passed in the midst of debate over the Wilderness Act, still 4 years shy of enactment. Regardless, Congress specified that wilderness was consistent with its intent in the Multiple-Use Act. Congress instructed the Forest Service to manage "all the various renewable surface resources of the national forests so that they are utilized in the combination that will best meet the needs of the American people" (16 U.S.C. 532). It further required that this management be "harmonious and coordinated . . . without impairment of the productivity of the land, with consideration being given to the relative values of the various resources, and not necessarily the combination of uses that will give the greatest dollar return or the greatest unit output" (16 U.S.C. 531). As one federal district court judge recently commented, the Multiple-Use Sustained-Yield Act "breathes discretion at every pore" (608 F.2d 803, 806).

Rather than bounding the decisions to be made, these mandates only increased the number of objectives to be satisfied by land managers with no direction given as to how the inevitable conflict between objectives should be resolved; how can these decisions be made "harmonious" when what is at stake is so great and when decisions clearly benefit some at the expense of others? How would the conflict generated between different groups competing for use of the same lands be resolved so that decisions will be harmonious? Congress apparently thought that the answers to these questions would come from involving the public in decisions and from making the decision-making process a rational one.

To ensure that land managers consider the consequences of their decisions on all facets of natural resource use, the Forest Service must comply with the National Environmental Policy Act (NEPA). NEPA requires that federal agencies complete environmental impact statements for any decisions that significantly affect the quality of the human environment. These statements must include descriptions of the unavoidable adverse impacts of the decision, alternatives to the decision, the short-term versus long-term consequences for productivity, and "any irreversible and irretrievable commitments of resources" resulting from the decision (42 U.S.C. 4332). However, like the substantive objectives of the natural-resource-management statutes, completing these statements is also a subjective process. Who defines what a "reasonable alternative" is? When is an impact adverse and when is it actually a benefit?

Lawrence Bacow (1980) has discussed the inherent difficulties in fulfilling NEPA's requirements: Which alternatives should be selected from the potentially endless list? How should public officials assess impacts when subjective judgments about risk and acceptability must be made? How and when should public input be obtained when problems are frequently too complex for public understanding? Finally, how might environmental and other interests be "balanced" when all are legitimate? Bacow (1980) argues that the premise NEPA is based on— "that the 'right' information is out there waiting to be gathered and, once collected, it will help us find the 'right' solution" (p. 122)—is invalid. He concludes that: "The fallacy in this argument is that the source of conflict is over facts when it is usually a difference in beliefs or values" (p. 122). Most of these decisions require value judgments about what is at stake and what is to be gained. Once again, technical expertise provides limited direction in making such decisions.

In addition, in the Forest and Rangeland Renewable Resource Planning Act of 1974 (as amended by the National Forest Management Act of 1976), Congress required that comprehensive land-use plans be developed and periodically revised for the National Forest System. Moreover, these plans must be "considered with the land and resource management planning processes of state and local and other federal agencies" (16 U.S.C. 1604). In fulfilling this mandate, Congress called for a "systematic interdisciplinary approach to achieve the integrated consideration of physical, biological, economic and other sciences" (16 U.S.C. 1604). Consequently, the USFS must now coordinate all timber- and fuels-management decisions with National Forest plans to help guide decision-making even though the plans themselves invariably involve value judgments about appropriate uses.

When Senator Hubert H. Humphrey introduced the National Forest Management Act to the Senate floor, he emphasized that "this bill is designed to get the practice of forestry out of the courts and back to the forests" (Hall and Wasserstrom, 1978, p. 523). But this comprehensive planning effort and coordination with routine administrative decisions has not made the land managers task any easier. Choices still must be made between different objectives. As seen in Chapter 7, it has only increased the number of occasions when Forest Service officials must make value judgments and, thereby, the number of occasions these judgments can be called into question.

MINERAL LEASING RECONFIRMED

Just as the environmental activism of the 1960s and early 1970s produced many environmental protection laws, an energy crisis and a faltering economy in the late 1970s and early 1980s created statutes and a political climate favoring precisely the reverse. As might be predicted, Congress responded to these new problems in like manner. Confronting an industry backlash at its earlier environmental excesses, Congress passed new legislation reinforcing the importance of mineral and fuels production from public lands. Two acts were passed in 1980, both hedging previous environmental mandates without actually rescinding them.

In the National Materials and Minerals Policy Research and Development Act of 1980 (30 U.S.C. 1601–1605), Congress reinforced the policies established in the mineral leasing laws. It instructed the President to "encourage Federal agencies to facilitate availability and development of domestic resources to meet critical materials needs" (30 U.S.C. 1601). Like past mandates, though, this development was not to be encouraged at the expense of all other resource values. Congress made it clear that "appropriate attention" was to be given "to a long-term balance between resource production, energy use, a healthy environment, natural resource conservation, and social needs." Congress was more direct in the Energy Security Act of 1980. It specifically instructed the Forest Service to promptly process all oil and gas lease and permit applications forwarded to it by the BLM or USGS. The agency could no longer defer these decisions pending completion of its long-term planning processes mandated by the National Forest Management Act of 1976:

> It is the intent of Congress that the Secretary of Agriculture shall process applications for leases on National Forest System lands and for permits to explore, drill and develop resources on lands leased for the Forest Service, notwithstanding the current status of any plan being prepared . . . (42 U.S.C. 8855)

In these acts, Congress once again reinforced the extent to which natural resource objectives compete and once again shifted the tough decisions to the land management agencies.

MAKING DECISIONS WHEN OBJECTIVES CONFLICT: THE BOB MARSHALL CASE

As seen in the cases discussed in Chapter 4, the land management paradigm that directed decision-making in earlier times, fails when these more recent objectives must be accommodated. The clash of the conservation and preservation philosophies and the inability of the paradigm to reconcile them is well illustrated by the following dispute over seismic testing in the Bob Marshall Wilderness Area.

The Bob Marshall ecosystem straddles 150 miles of the continental divide in Montana. At its heart is the 1-million acre Bob Marshall Wilderness Area, officially designated Wilderness in 1964 when the Wilderness Act was enacted. To its north lies the 286,700-acre Great Bear Wilderness (designated in 1978) and to its south is the 240,000-acre Lincoln-Scapegoat Wilderness (designated in 1972). The Bob Marshall Wilderness has been described by preservationists as "the quintessence of wilderness" (*High Country News,* December 28, 1979, pp. 4–5). Environmental groups following the wilderness-designation process believe it to be the crown jewel of the National Wilderness Preservation System (*High Country News,* September 18, 1981).

In early 1979, Consolidated Georex Geophysics (CGG), a Denver, Colorado, geophysical exploration firm, applied to the USFS Northern Region for a prospecting permit to conduct seismic testing in the Bob Marshall, Great Bear, and Lincoln-Scapegoat Wilderness Areas and on national forest lands adjacent to these wilderness areas. CGG's plans included detonation of 270,000 pounds of explosives along a 207-mile seismic line. All access along the line would be by helicopter and the testing was planned to last approximately 100 days (*High Country News,* May 1, 1981). At the same time as CGG's application was filed, oil and gas lease applications for the three wilderness areas began pouring into the district forester's office, eventually numbering 700 by 1982 (Findley, 1982, p. 311).

Announcement of the proposed seismic testing generated immediate outrage from both local and national environmental organizations. The Bob Marshall Alliance, a coalition of local individuals and regional and national environmental organizations, was established to oppose any oil and gas exploration or development intrusion into the wilderness areas. They referred to the proposed seismic testing as "bombing the Bob" and exclaimed that it would be "like slashing the face of the Mona Lisa" (*High Country News,* December 28, 1979, pp. 4–5).

The proposed seismic study area sat atop the Western Overthrust Belt and thus could prove to be a major source of oil and gas. It was also a pristine wilderness area, though, harboring one of the last remaining endangered Grizzly Bear habitats. Facing this dilemma, Regional Forester Charles Coston chose to make a decision based on a technicality. In so doing, he did not have to decide which objective (resource preservation or domestic energy production) should prevail in the area. CGG did not possess any leases in the Wilderness areas, and it intended to sell the information obtained from seismic testing to several oil and gas companies interested in obtaining leases there. Because CGG did not possess the leases, Coston ruled in April, 1980:

> It is my policy not to consider any proposal for mineral activities within a national forest wilderness unless the proposal is being specifically applied for under the United States mining laws or law pertaining to mineral leasing. (*High Country News*, May 2, 1980, p. 7)

Coston cited conflicts among "competing interests of multiple resources" in making his decision. He justified the decision, commenting that "it seems only prudent when discretion rests with the agency, that the conflict should be avoided" (*High Country News*, May 2, 1980, p. 7).

Coston's decision, however, hardly "avoided" conflict. Although environmentalists were jubilant, praising Coston's action, CGG appealed the Regional Forester's decision to USFS Chief Max Peterson. CGG was supported in its appeal by the Mountain States Legal Foundation (MSLF), an industry interest group, and the Rocky Mountain Oil and Gas Association (RMOGA) (*High Country News*, April 3, 1981). The Sierra Club, The Wilderness Society and the Bob Marshall Alliance immediately submitted their own briefs to USFS Chief Peterson supporting Coston's decision.

One year later, in April, 1981, USFS Chief Peterson sent CGG's application back to Regional Forester Coston for reconsideration. He cited the USFS responsibility under the Energy and Minerals Security Act, as well as the Wilderness Act, to not interfere with the operation of the mineral leasing laws. He ruled that Coston's decision to deny CGG's application because it lacked leases to the area was inappropriate. Peterson told Coston that "the citizens of the United States have an interest in assessing all values lying in those wildernesses" (*High Country News*, April 3, 1981, p. 2).

Now Coston was forced to make a decision on the merits of CGG's proposal. In late May, 1981, the regional forester again denied CGG's permit, this time ruling that "seismic operations conflict with significant wildlife, geologic, scenic, and recreation values" (*High Country News*, May 29, 1981, p. 14). Again, environmentalists were jubilant. Again, CGG, MSLF, and RMOGA began preparing administrative appeals of the Regional Forester's decision as well as lawsuits against the agency (*High Country News*, June 12, 1981, p. 2). And, once again, the Regional Forester's decision hardly put the issue to rest. While CGG, MSLF, and RMOGA prepared new administrative appeals and laid plans for a lawsuit against the USFS, environmentalists pressured Congress to take action to protect the Bob Marshall Wilderness (*High Country News*, May 29, 1981, p. 14).

CGG's proposal placed Regional Forester Coston in a corner. His position gave him responsibility for preserving wilderness values as well as fulfilling the provisions of the mineral leasing laws. In this case, however, it appeared impossible to accomplish both. Regardless, Coston had to make a decision. Whichever objective he fulfilled, he was doomed to encounter opposition. No decision was obviously correct; any decision could be supported as well as opposed on the basis of federal natural resource and land management laws. Coston's training as a professional forester did not help him find a path out of this corner. He had to exercise his judgment as to which resource values should prevail. CGG, MSLF and RMOGA assessed this situation differently and took Coston to court.

WHY THE CONGRESSIONAL INCONSISTENCIES?

As seen, Congress has delegated broad responsibilities and considerable decision-making discretion to professional land managers. This legislation implicitly assumes that the technical expertise of the land managers gives them the ability to establish the "relative values of the various resources" and thus equips them to reach decisions that "best meet the needs of the American people" (Hall, 1963, p. 287). In so doing, Congress reconfirmed the paradigm established consistent with the conservation ideal, while in the same voice undermining the ideal itself.

"Relative values" differ for different groups and individuals. Technical expertise alone does not lead to a decision; it must be combined with value judgments. As a result, in making oil and gas leasing and permitting decisions, the USFS must choose which objective to satisfy. A Forest Service decision supporting mineral exploration or development, for example, might completely undermine an endangered species protection or wilderness preservation objective. In like manner, a decision to prohibit oil and gas activity in a specific area runs contrary to expressed legislative and executive objectives to promote domestic energy production. The professional judgment exercised by the Forest Service in making such choices can easily be questioned by groups who value the resources at stake differently and who thereby legitimately reach different conclusions.

Although many blame Congress for the current stalemate in land management, Congress in reality had little choice but to act as it did. As a representative body, it is attuned to the demands of many different constituent groups. The difficult social-choice problems placed on its lap in the 1960s and 1970s could only be resolved through compromises that tried to accommodate all concerns. In so doing, Congress was able to garner support, rather than opposition, for its decisions. Members of Congress last but short terms if they ignore their constituencies and support unfavorable legislation. Unfortunately, as seen, Congress's responsiveness only compounded the difficulty of the land managers task. By legitimizing almost every conceivable public-land use without clear guidelines for choosing which use takes priority in a particular case, the tough land-use allocation decisions were tossed back to the Forest Service like a hot potato.

The symbolic dimension of these Congressional mandates cannot be ignored (Yaffee, 1982). There is little cost to most legislators in supporting broad policies regarding wilderness designation, pollution control, endangered species preservation, and watershed protection, or in making broad, seemingly reasonable demands for comprehensive planning of forest resources. At times, in fact, depending on prevailing public opinion, there is much to be gained by support of policies such as these.

Symbolic gestures by Congress are not the only reason these natural resource objectives are vague and often conflicting. There are real constraints on Congressional action on specific issues, especially as complex as those involving the public lands (Stewart, 1975, pp. 1667, 1695). It would be infeasible to expect Congress to specifically decide which wilderness or river is preserved, which timber sale is offered or which endangered species is protected. The time, resources, and information needed to make these decisions would only burden an already-overloaded Congressional agenda. Thus, our system of government is structured with Congress setting policy and the bureaucracy then implementing that policy.

Additionally, there are real political costs to making these decisions Congressionally, even if Congress had the time and resources. Because implementation occurs at the administrative, and not legislative, level, Congress does not have to make site-specific decisions in which the bottom-line costs and benefits of achieving its broad policy objectives are exposed for all to see. Instead, this latter side of the equation is passed on to the administrative agencies and it is there that the tough decisions must be made. The consequence is that Congress's broad policies have not resolved the disputes between competing groups; they merely shifted them back to the administrative arena wherein they often originated.

Because of these Congressional mandates, the Forest Service finds itself confronting many competing policy objectives. And, in each site-specific case, it must make the political resource allocation decisions that

Congress was either unable or unwilling to make. Administrative implementation of Congressional mandates is inherently problematic (Bardach, 1977; Edwards, 1980; Pressman and Wildavsky, 1973; Sabatier and Mazmanian, 1984; Yaffee, 1982). Bureaucratic inaction or purposeful misinterpretation of Congressional mandates frequently undermines policies. Such inaction is seldom a problem in natural resource decision-making, though, for those groups that fought for the legislation in the first place now fight to ensure that it is implemented as they intended. Their voices and frequent lawsuits do not allow an administrative agency to overlook or ignore those statutory provisions that advance their concerns. As Steven Yaffee (1982) discovered in his analysis of the implementation of the Federal Endangered Species Act:

> Constituent support is one of the most effective initiators of external pressures to force the bureaucracy into action. These groups . . . petition; they provide data; they educate; they lobby; they threaten legal action. Their actions account for many of the [endandered species] listings that were finally made. (p. 134)

Because the broad national resource mandates have legitimized their claims, it has given public-land interest groups power with which to question administrative decisions in the courts. This distribution of power, combined with the willingness of the courts to hear these cases, has dramatically changed the environment within which current land-management decisions are made. Just as in the administrative arena, however, the courts have had difficulty responding to the competing claims of equally legitimate parties.

THE LIMITS AND CONSEQUENCES OF JUDICIAL REVIEW

In 1977, the Forest Service mistakenly issued leases for a wilderness area within the White River National Forest. As a result, no protective stipulations were attached to the leases. When the error was discovered, Forest Service officials had several options, none of which were attractive. The Regional Forester decided not to do anything to rectify the error, reasoning that the agency would be sued regardless of what it did:

> The possibilities as we saw them: we may be sued for the error of agreeing to lease without special wilderness character protection stipulations: we may be sued for permitting operations on the lease without a full EIS; we may be sued for agreeing to lease even with protective stipulations; or we may be sued for not honoring a contractual obligation of the United States. I chose to honor the contract. (Nelson, 1982, p. 19)

Increasingly, the Forest Service is finding itself defending its decisions in court. During and before the 1960s, the Forest Service never had more than one, possibly two, court cases questioning administrative decisions pending at a single time. Today a forestry attorney in the Department of Agriculture's Office of the General Counsel puts that figure between one and two dozen (Brizee, 1975, p. 424). This outcome of decision-making occurs so frequently that agency officials are resigned to its inevitability. Because of the broad legislative mandates under which the agency operates, almost every decision made is susceptible to judicial review by an adversely affected individual or group.

On the one hand, whereas there has been a decreasing confidence in the Forest Service and, indeed, administrative agencies in general, on the other hand there has been an increased reliance on the courts to review contested administrative decisions. As U.S. Circuit Court Judge Harold Leventhal commented, the courts have come to share "the public sense of urgency reflected in the new [environmental] laws" (Altshuler & Curry, 1975, p. 31). More generally, in his review of the growing role and consequences of courts in social policymaking, Donald Horowitz (1977) observed:

> American courts have been more open to new challenges, more willing to take on new tasks. This has encouraged others to push problems their way—so much so that no courts anywhere have greater responsibility for making public policy than the courts of the United States. (p. 3)

Thus, it is not surprising to see the success with which public land users have taken their cases to the courts in hopes of redressing administrative wrongs. What better place to take one's problems than to a "judge" who, by definition, will address the fairness issues allegedly ignored by the administrative branch. As Horowitz observed:

> There is an undeniable attractiveness to the judicial method. In its pristine form, the adversary process puts all the arguments before the decision-maker in a setting in which he must act. The judge must decide the case and justify his decision by reference to evidence and reasoning. In the other branches, it is relatively easy to stop a decision from being made—they often effectively say no by saying nothing. In the judicial process, questions get answers. It is difficult to prevent a judicial decision. No other public or private institution is bound to be so responsive. (p. 22)

Although such reasoning places the judiciary on a pedestal of responsiveness, it presents only a fraction of a much larger picture. Horowitz goes on to note that:

> The fact that there are fewer participants in the adjudicative process than in the legislative process makes it easier for judges than for legislators to cut through the problem to a resolution. But it is precisely this ability to simplify

> the issues and to exclude interested participants that may put the judges in
> danger of fostering reductionist solutions. (p. 23)

Because judges are generalists and because issues are necessarily sim-
plified, the courts are not always the most appropriate forum for resolving some social policy disputes.

In analyzing several court cases involving social policy, Horowitz
found that judicial review was inadequate along several dimensions.
First, adjudication is focused. It is centered on the rights or duties of
different parties and therefore ignores discussion of alternative actions
wherein resolution of the underlying disputes might be achieved. Second, it is piecemeal. It deals only with the case before it. Third, it only
takes action when cases are brought before it. In this sense, it is a
passive body; it does not initiate action when needed. Fourth, adjudication focuses on historical fact or "events that have transpired between
the parties to a lawsuit," not social fact (Horowitz, 1977, p. 45). Social
facts, however, are those most often relevant to policymaking. Finally,
adjudication provides no room for planning. By nature, "judges base
their decisions on antecedent facts, on behavior that antedates the litigation" (p. 51). In social policy, though, it is critical to plan for the future
given the knowledge garnered from past experience. As a result, the
courts address the procedural, adjudicative issues, but the larger, substantive policy issues are frequently sidestepped.

In the case of public land management, the courts are encountering
much the same problem as Congress—how to choose between legitimate yet competing claims to public land resources. Just as there is no
obviously correct decision to the land manager, so too there is no obviously "fair" decision to the judge. It is a question of "right versus
right." Judges, like administrators, are also faced with Congress's obviously conflicting mandates when ruling on a case. Because legislated
objectives clearly conflict, some analysts argue that no Congressional
intent actually exists:

> the new areas of [judicial] activity respond to . . . new legislation so broad, so
> vague, so indeterminate, as to pass the problem to the courts. They then have
> to deal with the inevitable litigation to determine the "intent of Congress,"
> which, in such statutes, is of course nonexistent. (p. 5)

With such flexibility in determining Congressional intent, it is only to be
expected that many different perspectives have been offered.

When the Izaak Walton League sued the Forest Service to prevent
mining in northern Minnesota's Boundary Waters Canoe Area (BWCA),
Minnesota's U.S. District Court Judge Philip Neville decided to issue a
permanent injunction against future mining and mineral exploration

there (Wright, 1974, pp. 21–31). The BWCA, a part of the Superior National Forest, was designated as Wilderness when the Wilderness Act was passed in 1964. Judge Neville interpreted Congress's intent in this manner:

> To create wilderness and in the same breath to allow for its destruction could not have been the real Congressional intent, and a court should not construe or presume an act of Congress to be meaningless if an alternative analysis is possible. (Frome, 1974a, p. 144)

He concluded that:

> It is clear that wilderness and mining are incompatible . . . If the premise is accepted that mining activities and wilderness are opposing values and are anathema each to the other, then it would seem that in enacting the Wilderness Act, Congress engaged in an exercise of futility if the court is to adopt the view that mineral rights prevail over wilderness objectives. Mineral development by its very definition cannot take place in a wilderness area (*Casper Star Tribune*, November 15, 1981, p. A9)

But, encountering a similar question of whether or not the Forest Service should defer leasing decisions until wilderness designation decisions were made, Wyoming District Court Judge Clarence Brimmer issued a ruling directly conflicting with Judge Neville's. In *Mountain States Legal Foundation* v. *Andrus*, Brimmer ruled that decisions on the leases could not be deferred:

> It would surely be inconsistent with the intent to keep lands designated as wilderness areas open until December 31, 1983, that lands merely under administrative study for a proposed wilderness area may be effectively withdrawn from the leasing without the consent of Congress. (499 F.Supp. 393)

Brimmer commented that it is the Secretary of the Interior's obligation "to provide some incentive for, and to promote the development of oil and gas deposits in *all* publicly-owned lands of the United States through private enterprise" (499 F.Supp. 392). In so doing, he interpreted Congressional intent in the Wilderness Act's Section 4(d)2 in precisely the opposite manner as Judge Neville.

In contrast, in a case involving a timber sale on national forest lands *adjacent* to a proposed wilderness area but itself possessing wilderness characteristics, a Colorado District Court judge ruled that no commercial development that might harm the area's wilderness attributes could be allowed until after final wilderness decisions had been made by Congress. He ruled that the Forest Service must "refrain from acts . . . which will irrevocably change [the area's] character until the President and Congress have . . . rendered a decision" (309 F.Supp. 600). The result of cases such as these has been the development of a case law that is internally inconsistent and that only shifts the dispute back to the administrative agencies. The inherent conflict over the specific allocation and management of national forest lands is seldom resolved.

As anyone, judges prefer to have some substantive basis for ruling in a case. However, the extent to which Congressional objectives conflict forces the courts to focus on procedural issues and not question the professional judgment exercised by the administrative agencies. For example, the U.S. District Court Judge ruling in *Sierra Club* v. *Hardin* expressed the court's hesitation to question the substance of a decision made by professional foresters:

> Congress has given no indication as to the weight to be assigned each value and it must be assumed that the decisions as to the proper mix of uses within any particular area is left to the sound discretion and expertise of the Forest Service. . . . Having investigated the framework in which the decision was made, the court is forbidden to go further and substitute its decision in a discretionary matter for that of the Secretary. (325 F.Supp. 99)

In a separate case (*Sierra Club* v. *Block*) in which a proposed road providing access to future national forest timber sales was contested, the judge ruled:

> The court may not and shall not substitute its judgment for that of the Forest Service as to the environmental consequences of the Forest Service's actions so long as the proper procedures were followed and the Forest Service considered the environmental consequences of its acts.
>
> Whether the Forest Service's choice to harvest the forest or to build this road in a magnificent wilderness area is a good one is not for this court to decide. (*Environmental Law Reporter, 13*, p. 20178)

In yet another case (*National Wildlife Federation* v. *U.S. Forest Service*) involving a contested timber sale, the U.S. District Court judge ruled that:

> The Forest Service has special expertise in this area, and its determination of what is silviculturally essential is entitled to great weight. (*Environmental Law Reporter, 14*, p. 20758)

Although judges obviously have little choice but to rule in this manner, such decisions unfortunately only shift the disputes back to the administrative arena where they continue.

Because most court rulings are made on a single procedural issue, contradictory rulings now complicate administrative implementation of natural resource statutes and, moreover, leave the door open for further judicial review of the decisions made. In national forest management, the procedural issues raised generally involve administrative responsibility under the National Environmental Policy Act and, more specifically, the timing and scope of environmental impact statements. In oil and gas cases, a second category of inquiry questions the authority of the Secretary to impose different conditions on leases issued.

In *Sierra Club* v. *Hathaway* the Sierra Club argued that the land managers must complete an EIS before issuing leases for geothermal resources. The judge, however, upheld the government's position. He ruled that no development could occur under the leases without future decisions by the administrative agency and therefore that an EIS should be prepared at that later time:

> the lessee could not proceed beyond casual use until a notice of intent or plan of operation had been *approved* . . . the leasing decision therefore did not represent a commitment of resources. (579 F.2d 1162)

This policy of deferring extensive environmental review until specific operating plans are submitted has been adopted by both the Forest Service and BLM. In a review of onshore oil and gas leasing and exploration, the U.S. General Accounting Office (GAO) (1981) reported:

> With respect to oil and gas, both the Forest Service and the Bureau of Land Management have adopted practices which allow them to continue processing leases without committing themselves to approval of future development operations. They incorporate stipulations . . . into some leases which give the agency control over whether development will eventually be allowed under the lease. (p. 23)

However, in a suit (*Conner* v. *Burford*) brought by environmental groups protesting a Forest Service decision to issue oil and gas leases without an EIS in Montana's Gallatin and Flathead National Forests, U.S. District Court Judge Paul Hatfield ruled that the agency's decision to forego an EIS was unreasonable:

> In this case, the leasing stage is the first stage of a number of successive steps which clearly meet the "significant effect" criterion to trigger an EIS. The potential for a piecemeal invasion of habitat is present when the agency is allowed to lease without a comprehensive analysis of all stages of oil and gas development. (*Environmental Law Reporter*, 15, p. 20608)

The Forest Service, like most agencies, is reluctant to devote too much time and resources to upfront environmental analyses on unknown or speculative projects before specific development proposals are before them. For example, the Regional Forester in the Palisades Further Planning Area leasing dispute justified his decision to issue leases without an EIS because of the uncertainty involved:

> Mineral deposits are unique and completely unknown prior to exploration and discovery. To assume that a NEPA document will or can explicitly define

all the unknowns and uncertainties of an undiscovered mineral deposit is unrealistic. The Palisades FPA-EA fulfills the spirit and intent of NEPA and the selected alternative allows for at least two additional NEPA actions should use and occupancy be required within the FPA. (Sirmon, 1980, p. 2)

In *Environmental Defense Fund* v. *Andrus,* however, the judge ruled that such uncertainty was precisely the reason EISs are required of federal agencies:

Section 102(c)'s requirement that the agency describe the anticipated environmental effects of proposed action is subject to a rule of reason. The agency need not foresee the unforeseeable, but by the same token neither can it avoid drafting an impact statement simply because describing the environmental effects of and alternatives to particular agency action involves some degree of forecasting. And one of the functions of a NEPA statement is to indicate the extent to which environmental effects are essentially unknown. It must be remembered that the basic thrust of an agency's responsibilities under NEPA is to predict the environmental effects of proposed action before the action is taken and those effects fully known. Reasonable forecasting and speculation is thus implicit in NEPA. . . . (596 F.2d 1092)

A similar debate has clouded Forest Service timber management activities over the past decade. In fact, Judge Hatfield's decision discussed above is based, in large part, on an appellate court's ruling over a proposed road in Idaho's Nez Perce National Forest. In that case, the judge ruled that before the road could be constructed an environmental impact statement must be prepared that analyzed the cumulative impacts of constructing the road and the eventual timber harvests. Past rulings had, at most, required EISs on the proposed roads, leaving aside future timber sales to later environmental analyses.

Because of the uncertainty inherent in its decision-making, Forest Service officials rely on stipulations on its leases and permits to protect resources until later points in the decision-making process when more complete information is available about a specific project and its impacts. The effectiveness and legality of these stipulations has itself been questioned, however. In *Conner* v. *Burford,* Judge Hatfield ruled that:

the issuance of a lease with an NSO [no surface occupancy] stipulation does not guarantee an EIS before any development would occur. In fact, NSO stipulations can be modified or removed without an EIS. (*Environmental Law Reporter,* 15, pp. 20608–20609)

The Rocky Mountain Oil and Gas Association questioned the legality of similar stipulations when suing the Department of the Interior over BLM wilderness protection stipulations applied to oil and gas leases in potential wilderness areas. U.S. District Court Judge Kerr interpreted the BLM guidelines and Wilderness Protection Stipulation (WPS) to mean that

> actions which have even a remote possibility of impairing an area's suitability
> for wilderness designation are not allowable because the agency cannot per-
> mit the possible wilderness characteristics to be destroyed before those char-
> acteristics have been determined to exist. (*Oil and Gas Journal*, November 24,
> 1980, p. 58)

Kerr referred to leases issued under such procedures as "shell" leases
because companies holding them would be prohibited from conducting
any exploration under them:

> The government argues that the inclusion of the WPS with the lease informs
> the lessee that development may or may not be allowed. Such an argument is
> a poor excuse for the end result. Once again a lessee could continue to pay
> rentals and not be allowed to develop oil and gas. . . . Such a system of
> issuing "shell leases" with no development rights is clearly an unconstitu-
> tional taking and is blatantly unfair to lessees. A lease without development
> rights is a mockery of the term lease. (500 F.Supp. 1345)

Judge Kerr concluded, ruling that the BLM must lease wilderness study
areas:

> Energy and resource development are critical to this country at this point in
> time, and the erroneous interpretation of the law as given by the Solicitor is
> not only clearly contrary to the law as enacted by Congress, it is also coun-
> terproductive to public interest. (*Oil and Gas Journal*, November 24, 1980, p.
> 60)

Other judges, however, have ruled in precisely the opposite manner. In the Palisades FPA case, *Sierra Club* v. *Peterson*, Washington D.C.
District Court Judge Aubrey Robinson upheld the Forest Service contention when he ruled that an EIS was not required because the stipulations
attached to these leases would sufficiently protect surface resources:

> The lessees may legally obligate themselves to lease conditions that may
> result in the inability to explore or develop; that is knowing risk the lessees
> wish to take. (*Jackson Hole Guide*, April 15, 1982, p. 1)

Judge Robinson termed Judge Kerr's earlier ruling in *RMOGA* v. *Andrus*
to be of "questionable validity."

Similarly, in *Alaska* v. *Andrus*, plaintiffs challenged an EIS covering
oil and gas drilling in the Gulf of Alaska. They argued that the alternative of attaching a termination clause to the leases had not been considered. The court agreed with the plaintiffs contention, arguing that
such a termination clause might provide "that the Secretary could terminate a lease if environmental hazards, unknown or unforeseen at the
time of leasing, subsequently arose or were discovered" (580 F.2d 465).
In so doing, the judge supported the argument that lease stipulations
are enforceable even if they prevent development. The Interior Board of

Land Appeals has ruled similarly in at least two cases (12 I.B.L.A. 382(1973); 78 Interior Decision 317(1971)).

But, in the Cache Creek and Little Granite Creek cases, Interior Solicitor Lowell Madsen rendered an opinion directly counter to those that support the enforceability of stipulations, no matter how stringent. Madsen ruled that a lessee has "the exclusive right and privilege to drill" and that if "the pending applications for drilling permits are denied, the Secretary will have exceeded the scope of his authority and will have rendered nugatory an inviolable right . . ." (October 10, 1980, Memorandum to John Matis, USGS Task Force Leader). Madsen's opinion critically altered the course of events in the Cache Creek and Little Granite Creek permitting cases discussed in Chapter 4. Following Madsen's ruling, the USGS/FS EIS did not contain a "no drill" alternative.

These contradictory court rulings and legal opinions have only served to exacerbate, not resolve, the many disputes involved in managing the national forests. Because of the range of different judicial opinions over land management activities, they provide little guidance to administrative agencies. Moreover, they encourage groups to strategically select those courts where they believe they will get a favorable ruling. Judicial review of these administrative decisions involves considerable time and resources on the part of all parties just to rehash old issues rather than focus on the substantive issues of concern. Moreover, to gain access to the courts these groups must usually raise a procedural question. Frequently, these procedural issues are not the same as the substantive issues that directly concern the plaintiffs. And, because the underlying disputes are not resolved, they merely flow into the next forum where they can again be debated.

Some judges acknowledge the underlying disputes in cases brought before them questioning the adequacy of agency environmental analyses. These "differences of opinion" are disregarded, however, because the agency is not deemed responsible for altering its plans or conducting additional analyses solely because their findings were disputed. The judge ruling in *Warm Springs Dam Task Force* v. *Gribble* found that:

> The relevant questions under the NEPA are whether such comments are made available to decision-makers, whether the differences of opinion are readily apparent, and whether they receive good faith attention from decision-makers. The purpose of an EIS is to disclose such issues, but the responsible agency is not under a duty to resolve them. (565 F.2d 549(1977))

Whether the responsible agency has a "duty" to resolve such disputes, however, is different from whether it is in the best interests of the agency to try to do so. As seen in Chapter 4, whether an agency is procedurally in the right may have little bearing on its ability to proceed with its plans should some groups and individuals feel that decision is inappropriate. For example, an unresolved leasing dispute inevitably arises again when an APD is filed. In fact, recently proposed drilling in the Mosquito Creek section of the Palisades area has led to much the same controversy and debate as seen earlier in Cache Creek.

CONCLUSION

The public-land management paradigm was developed to pursue conservation objectives. And, it succeeded in serving those ends for the first half of the twentieth century. Now, though, the traditional conservation objectives have been supplemented with preservation and noncommercial objectives. Inherent value differences separate these different objectives; scientific expertise and technical analysis cannot choose which objective should prevail in a particular case. However, because the long-held paradigm remains intact, that is precisely how these decisions are made today. The consequence, as has been seen, is persistent conflict. Professionals schooled and experienced in the conservation tradition are no longer trusted to make these choices by those adhering to preservation ideals. Moreover, these professionals are not capable of making these choices in a manner that accommodates all interests because of the differences in values involved. Because power has become well-distributed by Congressional mandates and favorable court rulings, land management is at an impasse in many cases.

Society has a penchant for leaning on scientists and experts for making the tough social choices that inevitably must be made, precisely because these decisions are difficult, controversial, and many outcomes are possible. Relying on technical analysis to make value judgments, however, is neither appropriate nor effective. When power becomes distributed, as it has over public lands issues, viable decisions are decisions that accommodate the concerns of all interests at stake to their satisfaction. As seen, technical analysis fails at this task. Congressional intervention and judicial rulings have similarly failed. Other means must be found.

The current debate over how public land decisions should be made is not new. Since the early 1960s, academics, practitioners, professional foresters and political scientists have studied and criticized the current process. Reform has been proposed and attempted. As discussed in Chapter 6, however, past analysts have blamed the agency and not the paradigm. As a result, efforts at reform have been misdirected and, thus, unsuccessful.

CHAPTER 6

Why the Problem Persists

The U.S. Forest Service task differs markedly from that encountered when the agency was established at the turn of the twentieth century and the mineral leasing laws were enacted in 1920. Professionally trained foresters now have resource-allocation responsibilities that defy their technical expertise and that are more overt than at any time in the past. These foresters are charged with meeting many contradictory objectives. Their decisions must be made in the face of uncertainty about potential outcomes. They are required to make decisions in "the public interest" yet no single public interest exists. Although the decision-making process is designed to make scientifically defensible decisions, the decisions to be made are inherently subjective; objectively correct answers simply do not exist.

The Forest Service now confronts a political resource-allocation task in addition to the traditional scientific land-management task to which it is accustomed. The decision-making process, however, remains one based in technical expertise. It provides no means for resolving the disputes that inevitably arise. It cloaks political problems in technical analysis. The past three chapters have described a process that is obviously mismatched to the problem it is meant to address. In fact, too obvious. Why does the mismatch exist and persist? Moreover, how might it be remedied?

The answer to the first question is complex. First, although not all past analysts (Andrews, 1979; Fairfax, 1981) have misread the nature of the problem posed by public land management, many have. Rather than addressing the scientific management paradigm within which the Forest Service makes decisions, these analysts have attacked the conservation ideal. By attacking the long-held conservation notion of good public-land management and by demanding that the agency consider other, noncommercial values in decision-making, these critics have recommended procedural changes aimed at the wrong target. It is the paradigm that needs adjustment; it is *how* these many conflicting values are considered that is critical to reform. As seen, interest "airing" is not the same as interest "accommodation." Within the scientific management paradigm, wherein experts acquire information and then make decisions, interest airing is all that occurs.

Second, as discussed in Chapter 5, Congress responded to the expanding demands over public land resources following World War II by demanding precisely what land management agencies are now doing: considering other values in decision-making. However, by shifting these tough social choice questions to the land management agencies, Congress implied (and at times explicitly stated) that these decisions were best left to scientific expertise. It implied that these decisions had right answers that could be determined through the rational, dispassionate analysis of scientific experts. In so doing, Congress reinforced a paradigm that fails to effectively guide Forest Service decision-making.

Finally, there are organizational limits to change in the manner I argue is needed. Whereas Forest Service officials acknowledge that a problem exists, they logically frame their solution in the context of the traditional paradigm. This response is due in part to the Congressional legitimization mentioned above. More critically, however. it is due to the culture of the public forestry profession that is imbedded in this paradigm. This chapter discusses both the prescriptive sources of failure and the nature of the changes implemented by the Forest Service response, in order to determine how the mismatch might be remedied.

PAST PERCEPTIONS OF THE PROBLEM AND PRESCRIPTIONS FOR REFORM

This study is hardly the first to critically analyze the Forest Service and to recommend reform. Since the early 1960s, the problem has been studied extensively by students of administrative law and behavior, foresters, academics, and groups and individuals with a stake in public land management. They have criticized the Forest Service's failure to recognize the limits of its expertise and to change its behavior and procedures accordingly. Whereas almost all have come to study land management because of the dissatisfaction of different user groups, none have discussed how the underlying conflict between these groups might be resolved. Whereas proposed reforms have become more exacting, they still are not well-suited to the problem at hand. Although the problem is often recognized as political, reform proposals are almost always directed at changing agency behavior within the context of the traditional professional-expertise land-management paradigm.

As the demands placed on national forests broadened following World War II, Congress responded by expanding the list of objectives to be met in national forest management. National policy no longer centered on conservation ideals but embodied preservation and noncommercial ends as well. Given the changed policy, the "problem" warranting the attention of past analysts became the inadequate representation of these newly legitimized values in national forest management.

Forest Service officials respond to these new demands in a professional manner. When making decisions, they "consider" the many resource values at stake in a particular proposal and then make decisions reflecting what they believe to be "the public interest." As might be expected, this determination reflects both their training in commercial forestry practices and the norms of the public forestry profession. Forest Service staff seem to have confidence in their ability to master this increasingly complex management task and still make decisions in the public interest. This confidence has several roots. It is partly born of the Forest Service's long tradition of decision-making in the public interest. It is also due to their systematic approach to decision-making, in which all issues are theoretically uncovered and considered before a decision is made. And, partly it derives from their belief that, even though current problems are much more complex than in the past, professional land managers are better able than other, potentially more political bodies to make these decisions; professional land managers understand what resources are at stake and have an eye for long-term efficiency. The apparent confidence of the Forest Service on this account, however, has not been universally shared. Some have argued that it is unethical for the agency to take the public interest into its own hands.

A QUESTION OF PROFESSIONAL ETHICS

In the mid-1960s, University of Montana forestry professor R. W. Behan was one of the first members of the profession to express uneasiness about the changing nature of the forest management task and the inability of professional foresters to respond to it. In an editorial in the *Journal of Forestry* (1966, p. 398), he criticized the ethic adhered to by the forestry profession, one articulated to Behan's freshman forestry class by "a forester of considerable professional status":

> We must have enough guts to stand up and tell the public how their [sic] land should be managed. As professional foresters, we know what's best for the land.

Behan referred to this ethic as "the myth of the omnipotent forester." In this editorial, he hoped to plant the seeds for reform. He emphasized the extent to which value judgments pervade forest decision-making and the need for professional foresters to acknowledge different values and consider them. He recognized the limits of professional land-management training to make the inevitable value choices:

> "Goodness" and "badness" in our society are collective value judgments, and land expertise is no better a qualification than many others for making them. (p. 400)

Additionally, he pleaded for a shift in forester loyalty from the profession and to society:

> For the "various ends of society" in our unique society, are and will be set only *by* that society, and not by a professional class of foresters. It is when we as professional foresters either can't or won't understand this that we get the most rapidly into the hottest water. (And our forestry school training helps us very little in sensibly avoiding getting there or capably getting out.) It is when we attempt to determine ends that "pressure groups" become the most hostile, challenging our leadership in resource conservation, and they do so quickly and properly. (p. 400)

He concluded that freshman forestry classes should not be indoctrinated with the myth of the omnipotent forester but rather should be told:

> We must have enough sense to stand up and listen to the public, and to work with it in setting forest land objectives. Then as professional foresters we can supply the technological means to these sociological ends, and not confuse the one with the other. (p. 401)

In the end, though, Behan offered no plan either for accomplishing the shift in attitude or determining what "society" wants. He defined the problem as one of an ethical bias on the part of the Forest Service, and thus believed that appealing to their sense of "right", combined with a more appropriate forestry education, would sufficiently amend the system.

Behan's prescription for expanding the realm of forestry education was echoed by others during the 1960s and 1970s. When Daniel Henning (1971) studied the wilderness designation process he found, not surprisingly:

> Wilderness preservation, in many cases, is dependent on the recognition of intangible and qualitative values as opposed to tangible or quantitative values as dollars or numbers of people. Yet federal agencies and many aspects

of society are committed to values of the first order (progress, materialism, tangibles) as contrasted to values of the second order (quality, intangibles, ecological considerations). (pp. 70–71)

He, too, blamed professional forestry education for this bias:

Forestry schools, professional and agency indoctrinations *are oriented to producing a timber-management orientation* in many governmental foresters. Although the claim of professionalism and agency objectivity is made by many resource managers and administrators in the wilderness classification process, value considerations pertaining to economics and mass recreation are obvious. (p. 72)

Henning called for opening the "closed ring" of forestry education that now only involves "technically trained forestry professors who are training students for government positions where the requirements are made by professional foresters." (1970, p. 137) He believed that a curriculum including courses dealing with values, ethics, and humanities would offset the apparent commercial orientation of professional foresters. He also believed that the use of committees and consultants with expertise in various issue areas could supplement the agency's expertise and help foresters "determine the social ends or value base of the public interest." (1970, p. 138) He denounced the current process in which "the same administrators who initiate and argue in support of a particular proposal" are the judges that then evaluate public input in reaching a decision. (1971, p. 72)

With time, university-level forestry programs have begun to offer or require courses on ethics and the value aspects inherent in a professional forestry career. In addition, the USFS has held its own workshops in interpersonal skills and different strategies for involving the public in decision-making. (U.S. Department of Agriculture, Forest Service, 1973) More recently, the Forest Service has begun sponsoring professionally run training sessions in conflict management and dispute resolution techniques for its field personnel (see Chapters 8 and 9). Although education in nontechnical forest-management considerations and professional training in conflict management and interpersonal skills are a critical first step in meeting the needs of today's land-management task, they cannot stand alone. At the bottom, education and open-mindedness do not give a professional forester the ability to actually represent the values of others in decision-making. Individuals and groups must represent their *own* concerns and values. The question facing the Forest Service is *how* to provide them an opportunity to do so; what process might successfully accommodate the concerns of affected user-groups to their satisfaction while retaining Forest Service authority over land management? Chapters 7–9 explore this question.

A QUESTION OF UNBOUNDED ADMINISTRATIVE DISCRETION

Although many have criticized professional forestry training as the source of inadequate interest representation in Forest Service decision-making, most have pinpointed the problem as one of unreined administrative discretion and an agency bias in favor of some interests over others. There have been two levels of attack on this account and thus two levels of reform prescribed. Some critics perceived a bias engrained in the agency that might be offset through education, public involvement in agency decision-making, hiring of professionals trained in areas other than forestry, and implementing more formalized procedures that would constrain these biases. Others perceived a bias that was imposed externally by agency "capture." All blamed Congress for the predicament.

Agency Bias

As early as 1962, Charles Reich asked many of the same questions that have been posed in this analysis:

> Management must decide between the competing demands on the forests. When different uses clash, which shall be favored? How are local needs to be balanced against broader interests? Who is to have the benefit of the economic resources, and on what terms? How are the conflicting recreational demands of fishermen, skiers, hunters, motorboat enthusiasts, and automobile sightseers to be satisfied? Should the requirements of the future outweigh the demands of today? (p. 1)

He criticized the power bestowed on "small professional groups. . . . [to] make bitterly controversial decisions, choices between basic values, with little or no outside check" (p. 2). He blamed Congress for delegating such extensive administrative discretion to the Forest Service with little attempt to bound the agency's authority:

> The standards Congress has used to delegate authority over the forests are so general, so sweeping, and so vague as to represent a turnover of virtually all responsibility. "Multiple-use" does establish that the forests cannot be used exclusively for one purpose, but beyond this it is little more than a phrase expressing the hope that all competing interests can somehow be satisfied and leaving the real decisions to others. The "relative values" of various resources are to be given "due consideration," but Congress has not indicated what those values are or what action shall be deemed "due consideration." Congress has directed "harmonious and coordinated management of the various resources," but it has left the Forest Service to deal with the problem that different uses of resources often clash rather than harmonize. Most significantly, Congress has told the Forest Service to "best meet the needs of the American people" but has left it entirely up to the Service to determine what those needs are. (p. 3)

Reich criticized Congress's "multiple-use" mandate in that it justified all land uses. In so doing, he charged that any Forest Service choice could be justified and few could be legally questioned.

Reich then turned his attention to how the Forest Service exercised its broad discretion. He lambasted the Forest Service presumption that it could determine what constituted the public interest before making a decision:

> the Service recognizes, in the matter-of-fact pages of its manual, that its ultimate job is nothing less than the definition of "the public good," a task once reserved for philosopher-kings. (p. 13)

Finally, he stressed that the Forest Service manner of decision-making inevitably led to bias in representing the many land-use values at stake:

> It is too much to expect that foresters who initiate and argue in support of a particular proposal can then adequately evaluate public criticism or counterproposals that often represent thinking they have earlier rejected. (p. 6)

Reich recommended three reforms. First, he called for public hearings on controversial proposals to ensure that the Forest Service was made aware of all concerns before making a decision. Second, he proposed that the many different resource values be institutionalized within the Forest Service by hiring staff other than foresters. In so doing, these other values would be automatically incorporated into day-to-day agency decision-making. Finally, Reich called for formal justifications when decisions are made so that all concerned groups know how and why the decision was reached. Reich believed that such justifications would make the Forest Service more accountable for their decisions and, ideally, force them to represent all interests in decision-making (pp. 7–11).

To some extent, all of Reich's reforms have since been institutionalized. As a result of National Environmental Policy Act procedures, as well as new agency policies described earlier, public hearings are held on most controversial decisions. Environmental assessments and environmental impact statements now contain the formal justifications he suggested. The agency also has expanded the profile of its staff in a special "public information office" as well as within its field staff (Cutler, 1972). But, as illustrated in earlier chapters, "all competing interests" have not been "satisfied" by these measures. Although most interest groups actively participate in formal agency hearings on various proposals and comment on draft analyses, they still do not feel that their concerns have been accommodated by the Forest Service decision-making process. Hence, opposition persists.

In the year following Reich's article, University of Virginia Economics Professor George Hall similarly blamed Congress for delegating broad discretionary authority to the Forest Service:

> It appears that in voting for multiple use Congress believed it was voting for virtue and against sin without having a definite idea about just what actions constituted virtue or sin. (1963, p. 282)

He criticized what he viewed to be the fundamental flaw of the multiple-use doctrine, that being that:

> it fails to provide a criterion for resolving the conflicts among demands except for the general and unspecified standard of "the best interest of the public."

Hall also criticized the apparent Forest Service bias in exercising its discretion in decision-making. He labeled a "myth," the Forest Service belief that its "professional competence . . . savvey [sic] . . . and good judgment" (p. 284) could successfully balance competing interests in multiple-use management:

> The myth is that such practices are capable of resolving all conflicting demands and allowing us to have our cake and eat it too. Resolution, of course, occurs in the sense that *some* land use plan is selected. (pp. 284, 289)

Hall's recommended solution to the multiple-use dilemma differed from Reich's. He proposed that procedures be instituted for assessing the costs and benefits of different land allocations. He believed that, in so doing, conflicting demands would be appropriately "resolved," thereby ensuring that the "socially best administrative decisions for the national forests were made" (p. 287). Hall also admitted the difficulty in devising such a scheme when so many intangible values were involved. Economists have often debated different means for quantifying environmental amenities and therefore remedying that failure of cost-benefit analysis. No specific measures have ever been generally accepted by these economists, though. (See, generally, Freeman, 1979). And, when the Forest Service has tried to quantify the values at stake in its decisions, it has, as seen, been soundly criticized. The groups involved seldom trust the agency's evaluation of their concerns and thus they oppose the outcomes of such analyses.

Reich's and Hall's prescriptions came before many natural resource protection statutes had passed, particularly the National Environmental Policy Act. As discussed in Chapter 5, these statutes eventually gave considerable power to user groups to oppose land-use plans and decisions deemed inappropriate. Thus, today's conflicting demands are *not* resolved when, as Hall implied, "*some* land use plan is selected" or decision is made (1963, p. 289). It was not until the early 1970s that the persistent and powerful opposition of dissatisfied user-groups became a cause for concern. At that time, analysts turned their attention to this consequence of the agency's broadened mission.

In 1972, M. Rupert Cutler analyzed, in his doctoral dissertation, the growing dissatisfaction with and litigation over Forest Service decisions. He attributed this trend to two factors. First, numerous court decisions in the late 1960s and early 1970s had liberalized judicial rules of standing in environmental disputes and thereby provided previously unavailable opportunitites for court review of Forest Service decisions. Second, he determined that these newly-opened doors to the judicial system were well-used because Forest Service decision-making procedures failed to provide adequate opportunities for public involvement.

Cutler recommended five specific reforms to overcome this problem of constant litigation, all designed to improve public involvement and Forest Service receptivity to nontraditional concerns. First, he called for an improved multidisciplinary profile within the agency. This profile could be established by "personnel recruitment and in-service training" and "two-way communication with clientele groups." Second, he encouraged early involvement of all "clientele groups" in agency decision-making. Third, he suggested "the use of independent hearing officers and semi-independent citizens' committees." Fourth, he called for the Forest Service to generate several alternatives for public review and comment and, finally, to provide enough time for different groups to conduct "adversary analyses" of agency actions (Cutler, 1972, pp. 498–509).

Many of Cutler's recommendations have been adopted to some extent by the Forest Service. "In-service" training efforts have been mentioned earlier. The multidisciplinary profile has improved, mostly through expanded hiring of land management professionals other than foresters. In 1959, foresters comprised 90% of the Forest Service staff as compared to 50% in 1975 (Robinson, 1975, p. 34). Extensive public workshops involving thousands of individuals were held during the Forest Service wilderness reviews (RARE I and RARE II) (U.S. Department of Agriculture, 1979). At times, "independent hearing officers" have been used to run public meetings. These meetings, however, were always to obtain public input, not resolve disputes. As a result, the independent hearing officers often only complicated the meetings by impeding the flow of information between Forest Service staff and those attending the meetings. Their efforts seldom facilitated discussion between the different groups (U.S. Department of Agriculture, Forest Service, 1973, pp. 44–49). "Semi-independent citizens committees" are seldom used by the Forest Service in decision-making. Red-tape associated with the Federal Advisory Committee Act is often given as a reason for not more formally involving such committees (U.S. Department of Agriculture, Forest Service, 1973; Wagar and Folkman, 1974, pp. 405–407).

This Forest Service response is partly due to Council on Environmental Quality regulations under the National Environmental Policy Act (*Federal Register*, November 29, 1978, Part IV, 55978–56007). Environmental impact statements must now contain analyses of several alternatives and public comment on a draft version of the report. But, to a large extent, today's public involvement efforts by the Forest Service are a result of Cutler's own efforts as Assistant Secretary for Conservation, Research, and Education in the Department of Agriculture during the Carter Administration.

Cutler was placed in the enviable position of implementing the recommendations made in his dissertation research. As Assistant Secretary, he immediately commented that the "one imprint" he wanted to make on the Forest Service was "development of policies that ensure extensive, high quality, public involvement in policy determination" (1978, p. 264). His reasoning was that:

> The amount of citizen litigation to block unacceptable decisions relates directly to the opportunities, or lack of opportunities, for public participation. The U.S. Department of Agriculture's aim is to make such litigation unnecessary, to tear down any remaining shrouds of administrative procedure that tend to shelter our decision-making, and to get everyone affected by the results of our decisions into the decisionmaking process." (1978, p. 264)

Cutler established five stages to be followed in decision-making in order to ensure successful public involvement (1978, pp. 264–265). Success, in this scheme, was reducing the extensive litigation plaguing the agency. These five stages were:

1. Defining the issue
2. Collecting public input
3. Systematic analysis
4. Evaluating public comments
5. Decision implementation

It was Cutler's belief, given his research findings, that more open and accessible decision-making would equate with representative decision-making. In theory, this representativeness would offset the widespread opposition to Forest Service decisions. The final test of this theory would be in the decision-implementation stage; to use Cutler's words, this stage would reveal "whether the public has been adequately involved in the decision" (1978, p. 265).

Cutler's proposals and the Forest Service response are to be commended. Dramatic change has occurred in the way the agency involves the public in decision-making (Culhane, 1981, Chapter 8; Dravnieks and Pitcher, 1982; Valfer, Laner and Moronne, 1977). If the "final test" is in decision implementation, however, these reforms have not been enough. The problem persists. Cutler's prescriptions have failed to end the litigation at which they were targeted because, like those before him, he prescribed reform that reinforced the scientific management paradigm and further entrenched the technical analysis approach to decision-making. Whereas the Forest Service has employed more and more sophisticated analysis techniques to systematically evaluate public input, the public is not satisfied and continues to try to redress the perceived wrongs of the administrative process by appealing to the courts.

Consider, for example, the showpiece of this participatory process: The RARE-II wilderness reviews. Fifty-thousand people were involved in providing input to the scope of the wilderness review EIS. Seventeen thousand more were involved in workshops to structure the review process. When the dEIS was released, 264,000 more comments were received (Cutler, 1978, p. 265). But, in "the final test," the process and its conclusions have been attacked on all fronts. Mining and timber industry groups, environmental organizations, ORV and other backcountry users have all opposed its conclusions as either too much or not enough wilderness protection (*The Living Wilderness*, January/March 1979, 42 [144]). And, moreover, because of a 1982 court ruling that this EIS was inadequate, the Forest Service is in the midst of an amended RARE-III process (*High Country News*, February 4, 1983, p. 2).

Whereas the problem addressed in this analysis closely parallels that studied by Cutler, the root of the problem as defined here is much different. This study argues that the paradigm in which technical analysis is used to determine outcomes is the root of the problem. Cutler's reforms reinforced this paradigm. Cutler admitted that "there is no formula in the decision process that tells USDA what weight public participation should receive relative to other factors" (1978, p. 266). Nonetheless, he still called for agency officials to "weigh public input against other decision factors" in the evaluation stage (p. 265). He believed that the public would have faith in the administrator's decision as long as public comments could be related to alternatives in a "consistent, visible and traceable way." This fishbowl decision-making effort, however, has not addressed what this analysis indicates is the heart of the problem: the many affected interests do not feel represented unless their concerns have been accommodated to their satisfaction. And, critically, they no longer trust the systematic, technical analysis of the public administrators to accomplish this end. Some other means for accommodating these many interests must be found if viable decisions are to be made. Systematic, technical analyses that do not conclusively pinpoint solutions should no longer be used to hide the judgments that invariably must be made.

Agency Capture

There was also a minority during this time that framed the problem somewhat differently than simply agency bias in decision-making. They argued that the Forest Service was "captured" by a single clientele group, and that other interests could never be represented under the existing system. Some claimed that the Forest Service was captured by environmental concerns (McNamara, 1979), whereas others believed that development interests were the culprit ("Managing Federal Lands," *Yale Law Review*, 1973). These analyses diverged from those discussed above and called for Congress to avert single-interest domination by setting the specific priorities to be adhered to by administrative agencies.

The *Yale Law Review* article analyzed the federal multiple-use management system and found that land management decisions benefit local and development interests at the expense of needed environmental protection. In a scathing attack of the multiple-use system, the author criticized the "vacuous platitudes" of the Multiple-Use and Sustained-Yield Act and called for a complete Congressional overhaul of the entire natural resource administrative system. In order to better represent environmental and other national interests, the author proposed three new agencies, each with a specific management task: grazing, timber, and recreation (including wilderness). In this scheme, Congress would reshuffle all public lands and then allocate them to one of the three categories. The specific resource value would then prevail, with other uses only allowed when compatible with the primary use. Congress would make these broad land allocation decisions with the help of special commissions consisting of "industry representatives, conservationists, and some agency specialists" (p. 801). The Commission's proposals would solely be advisory and Congress would have power to change allocations whenever national needs dictated. Current agency officials were to be kept out of this allocation process, however, because they "have a penchant for empire building, and are more likely to fight for expanded domains than to fairly articulate and choose among the interests involved." The author seems confident that this type of system would be better able than the existing system to end the persistent opposition to land management decisions:

Resort to adjudication would be much less frequent . . . The inefficiencies of using the cumbersome, case-by-case adjudicative process for policymaking would disappear once policy was established by Congress. Adjudication would become a process of testing an administrator's actions against Congressional directives, as it ought to be. Since there would be manageable directives from Congress greatly narrowing the range of discretion, appeals within the agency and to the courts could provide meaningful control. (p. 803)

But, it is not clear that the basis for this article's criticisms—that the Forest Service is captured by local and development interests—is well-founded. Paul Culhane (1981) questioned the periodic attacks leveled against the Forest Service by those adhering to the capture theory. In a comprehensive analysis of Forest Service decisions, he found that, far from being captured, Forest Service decisions appear, overall, to be quite well-balanced among the many different interests involved.

Regardless, such an extreme response, if not unnecessary, is politically infeasible. As discussed in Chapter 5, Congress mandated its long list of legitimate public-land uses for many reasons. It is seldom politically costly to support legislation that promotes many varied uses for the national forests as opposed to special, single-purpose uses. In fact, sometimes it is beneficial to do so as such legislation is often symbolic and enhances a legislator's reputation (Yaffee, 1982). Furthermore, the "vacuous platitudes" of many Congressional acts are a result of real constraints—both political and technical—on greater specificity (Stewart, 1975, pp. 1677, 1695). Congress would be overwhelmed if it tried to dictate precise measures for implementing its mandates. Making site-specific choices between, for example, energy development or endangered species protection or wilderness preservation would only further crowd Congress's already-burdened agenda. It would also be politically costly for many legislators to do so because it is at this site-specific implementation of Congressional mandates that the "winners" and "losers" become clearly defined. When possible, legislators prefer to avoid resolving such controversial choices (Stewart, 1975, p. 1677). Therefore, it is highly unlikely that Congress would ever voluntarily take on the hot potato suggested in the *Yale Law Review* article. And, in fact, they have refused to do so on several occasions in the past (Culhane, 1981, p. 50; Fairfax, 1981, p. 516–582).

THE PROBLEM OF ADMINISTRATIVE DISCRETION: A LARGER VIEW

The common theme underlying these past critiques is the extent to which Congress has delegated broad administrative discretion to the

Forest Service. Critics argue that the problem persists because the Forest Service exercises its discretion in an inappropriate manner. Therefore, many measures are proposed to redirect or curb agency discretion as a way of allowing full representation of *all* interests.

Framed as a problem of excessive administrative discretion and thus agency bias in decision-making makes the Forest Service situation appear little different from that of many administrative agencies. Students of administrative law and administrative behavior have long criticized Congressional delegation of broad authority to nonrepresentative agencies. Their proposed reforms echo those recommended for the Forest Service.

As Richard Stewart (1975) argues in "The Reformation of American Administrative Law," however, these reforms are seldom appropriate. He describes the role of administrative agencies to be a political one of "balancing" competing interests, not solely a technical one:

> The application of legislative directives requires the agency to reweigh and reconcile the often nebulous or conflicting policies behind the directives in the context of a particular factual situation with a particular constellation of affected interests. The required balancing of policies is an inherently discretionary, ultimately political procedure. (p. 1684)

But, Stewart notes that most reforms that are proposed rely on Congress setting strict standards and procedures for agency compliance. Such prescriptions assume that today's complex problems can be solved using explicit procedural formulas applied in a professional manner. In so doing, all interests theoretically are fairly represented because administrators are no longer able to exercise discretion in favor of particular client groups. Stewart argues, though, that most administrative tasks are not adaptable to these formalized procedures:

> the relatively steady economic growth since World War II . . . has allowed attention to be focused on the perplexing distributional questions of how the fruits of affluence are to be shared. Such choices clearly do not turn on technical issues that can safely be left to the experts. (p. 1684)

As a result, he labels the traditional model of administration as "an essentially negative instrument for checking governmental power" (p. 1687). Because the decisions are inherently political, no technical formula can ensure that all interests are represented in decision-making. By trying to do precisely that, though, the traditional model relies predominately on the judicial system to review administrative decisions after they are made. Stewart argues that a more appropriate approach would be an "affirmative" system of "government 'which has to do with the representation of individuals and interests' and the development of governmental policies on their behalf" (p. 1687).

Unfortunately Stewart does not recommend any means for developing this "affirmative" system, one recognizing the political dimensions of today's complex social choice problems. His conclusion is that general reform is impossible but instead that energies should be devoted "to treating various instances of administrative 'failure' through case-by-case examination" (p. 1805).

ONE CASE EXAMINED: NATIONAL FOREST MANAGEMENT

Unlike many administrative agencies, the Forest Service was not established to either serve or regulate a particular clientele. Instead, it was charged from the outset with no less a task than managing the national forests in "the public interest," and to serve "the greatest good of the greatest number in the long run." For much of Forest Service history, achieving this end was a relatively easy task. "The public interest" was perceived to be the outcome of professional, scientific land management. Hence, if the agency responded to its task professionally and managed the lands using scientific silvicultural, fire, flood, and pest control practices, the outcomes would, by definition, coincide with the public interest.

The conception of what constitutes "the public interest" has changed, however. No longer is the Forest Service merely able to exercise its technical expertise in land management. Now it is expected to also consider many nontechnical aspects of decision-making that are counter to the conservation values that dominated for so long. Now it is expected to weigh values that yield no measurable commercial return from the national forests. No longer do Forest Service scientific management practices and technical analyses automatically result in outcomes satisfying "the public interest." The agency is charged with making some difficult social choices that are political, not scientific, in nature. Like most administrative agencies, the Forest Service is unequipped to do so.

PUBLIC ADMINISTRATION AND THE PUBLIC INTEREST

The Forest Service is not alone in its charge to represent the public interest in decision-making while having many publics advocating just as many different public interests. That is the responsibility of most administrative agencies. The federal government generally adopts responsibility for tasks for which a broader public interest exists that otherwise would not be met. In fact, many administrative agencies arose out of the same scientific management ideal that gave birth to the Forest Service (Mosher, 1968). How do these other agencies define and then satisfy "the public interest"? As one student of public administration theory phrased it: "If the bureaucrat is exposed to the babel of many voices, speaking different tongues to convey various messages, how can he know which among competing alternatives he *ought* to choose?" (Schubert, 1957 p. 346).

Unfortunately, there is no single answer to this question. Public administration theorists have long debated but never resolved how administrators are to determine and then serve "the public interest." Three different theories have been offered regarding the role of public administrators: administrative rationalism, administrative platonism, and administrative realism. The Administrative Rationalists assume that efficient management coincides with the public interest. In Herbert Simon's words:

> The criterion which the administrator applies to factual problems is one of efficiency. . . . Once the system of values which is to govern an administrative choice has been specified, there is one and only one "best" decision. (Schubert, 1957, p. 349)

Although this view prevailed during the progressive conservation era when scientific management principles were devised, it is seldom professed today. Today the specified value system is internally inconsistent. For example, the problem frequently faced by the Forest Service is how to choose between commercial and noncommercial values that have legislatively been given equal weight even though they are often incompatible. The Rationalists' theory is unenlightening when, within the boundaries of "efficiency," several legitimate outcomes are possible.

The Administrative Platonists, though not in total agreement on how to operationalize their theories, view public administrators as taking a more active approach in defining the public interest and then using technical expertise to achieve it. For example, Paul Appleby notes that:

> Neither the simple reconciliation of private interests nor their reconciliation modified by considerations of public interest is in the end a technical performance, no matter how many technical factors may figure in it. It is a political function, involving essentially the weighing of forces and the subjective identification of the narrow area within which these forces may be balanced and the exercising of discretion concerning the point within that area at which acceptability and public interest may be effectively and properly maximized. (Schubert, 1957, p. 353)

To a large extent, the Forest Service adheres to this latter view of appropriate behavior. The agency acknowledges the many publics affected by its decisions and thus the necessity to move beyond just technical analysis in decision-making. Forest Service officials perceive it to be their responsibility to seek out and then weigh the public interest when making decisions. Unfortunately, though, this approach does not satisfy the many publics; the "babel of voices" only grows louder with their efforts. As one student of these theories concludes, the problem with the Administrative Platonists is that "they have described the public interest as a thing of substance, independent of the decisional process and absolute in its terms" (Schubert, 1957, 367).

The Administrative Realists, in contrast, argue that there is no such thing as "the public interest" and thus the key to administrative decision-making is in the process by which the "babel of voices" is quieted, not gelled into a single will. David B. Truman describes this view that "the public interest" is a nonexistant entity:

> Many . . . assume explicitly or implicitly that there is an interest of the nation as a whole, universally and invariably held and standing apart from and superior to those of the various groups included within it . . . such an assertion flies in the face of all that we know of the behavior of men in a complex society. Were it in fact true, not only the interest group but even the political party should properly be viewed as an abnormality. . . . Assertion of an inclusive "national" or "public interest" is an effective device in many . . . situations. . . . In themselves, these claims are part of the data of politics. However, they do not describe any actual or possible political situation within a complex modern nation. In developing a group interpretation of politics, therefore, we do not need to account for a totally inclusive interest, because one does not exist. (Schubert, 1957, p. 358)

The Realists' observation coincides with the current Forest Service dilemma. There are many publics affected by agency decisions and each is advocating decisions satisfying its particular concerns, arguing that they coincide with "the public interest." The Forest Service decision-makers have been unable to define a particular "public interest" to which the many publics will agree.

It appears, then, that the Realists might provide some direction for the Forest Service out of its current dilemma. One of the major criticisms leveled against the Realists, though, is that they have been unable to articulate an operational definition for converting their theory to practice. Perhaps the closest guide to implementation is Appleby's statement that public administration:

is the eighth political process. It is a popular process in which vast numbers of citizens participate, in which assemblages of citizens comprise power units contending with each other, in which various governmental organizations are themselves functional representatives of special interests of many citizens, and in which these organizations themselves contend mightily with each other in the course of working out a consensus that translates many special interests into some workable approximation of public interest. This process is as essential to the evolvement of governmental action as public debate, and closely akin to it. (Schubert, 1957, p. 358)

The Forest Service implicitly rejects the Administrative Realists' theory, though. Agency officials argue that it is their responsibility to make these decisions, not to turn their authority over to the vocal, but not-necessarily representative, public (Interview with former USFS Chief R. Max Peterson, October, 1982). Instead, the agency seems to adhere to the administrative platonist point of view. Agency officials seem confident that by determining and then analyzing the many different interests at stake, an outcome can be determined that will coincide with the public interest. However, the "babel of voices" that arises when these decisions are to be made is not quieted by administrative decisions stamped with the Forest Service label: "Made In The Public Interest." The Forest Service adherence to a platonist-type ideology does not guide it out of its predicament.

THE USFS DECISION-MAKING PROCESS

USFS officials are fully aware of the many perceptions of the public interest in their decisions. They, perhaps more than anyone, acknowledge the inherent difficulty of the decisions to be made in the face of these competing demands. In many ways, they feel vindicated by these contrary opinions. If left to its own devices, the public would be constantly embattled, never able to decide who gets what. By having an agency of trained public-land managers, these decisions can be made in a systematic and scientific manner in which all the different competing interests can be heard and their interests balanced. Theoretically, this approach should lead to decisions that best approximate "the public interest."

The *Forest Service Manual* clearly states that it is the Forest Service's administrative responsibility to "weigh the public's interest" in decision-making:

The Forest Service . . . encourages identification of the values to be protected, definition of resource protection and mitigation needs—based on relative values of minerals and impacted resources—and a weighing of the public interest where there is a conflict. If, after such weighing of public interest, the

> potential adverse impact of mineral development on surface values of an area
> is judged to be unjustified, considering the value of the minerals, the Forest
> Service shall recommend (or require, where appropriate) no leasing. (Section
> 2827.14b; Interim Directive No.6, December 31, 1980, p. 8)

Assessing the "relative values" of minerals, timber, wildlife, and other affected resources and "weighing the public interest" are much easier said than done, though, especially when there are so many publics, each advocating different outcomes, each allegedly representing "the public interest." How do Forest Service field staff implement this mandate?

The Cache Creek/Little Granite Creek drilling-permit review process illustrates the thorough information gathering and public participation efforts of the Forest Service in decision-making. It is a process being mirrored today in comprehensive forest planning processes in all 155 national forests. In June, 1980, a 3-day public scoping meeting was held in Jackson, Wyoming, to help the U.S. Forest Service/U.S. Geological Survey task force in charge of developing this EIS determine what issues and topics should be addressed in it. Before the public scoping meeting, the interagency task force drafted a preliminary scoping document to provide a base for discussion at the meeting. This document was distributed to over 300 groups, individuals and public agencies. Responses as well as additional ideas could be presented at the formal scoping meeting, in writing to the USGS or by calling the USGS collect. In addition, public hikes to both proposed wellsites augmented the scoping meeting (U.S. Department of Agriculture, Forest Service, 1981).

Public input and fact-finding were hardly limited to this 3-day scoping session, though. The USGS/FS task force prided itself in its extensive "consultation with others." In Chapter VIII of the draft EIS, the interagency task force discussed the tremendous amount of input and analysis supporting the EIS conclusions. The discussion noted that the original USFS Environmental Assessment on the proposal was completed in December, 1977, and that more than 200 written comments were received on it. A formal public meeting was then held in January, 1978. Next, an additional USGS "unusual" environmental assessment prompted more input both in writing and public hearings held in late July and early August 1979. The chapter went on to list the task force's information gathering efforts:

> Over 160 people attended the public (scoping) meeting, and some 35 individuals participated in the "show-me" trips to the proposed wellsites.
>
> Well over 1,000 private individuals were personally contacted or consulted for information on technical aspects by project scientists during the course of the EIS investigation.
>
> A series of [eighteen] reports was prepared by Government personnel and

private consultants to analyze and document various aspects of the Cache Creek-Bear Thrust EIS. The reports cover environmental, engineering and legal issues.

Personal contacts with concerned individuals were made by the task force leaders and other members of the USGS and FS throughout the analysis period. These personal contacts, and other informal meetings, were extremely important in developing an understanding of the various publics' positions. . . . All such information has been used to assess the issues and concerns, public needs, demands and alternative formulation (U.S. Department of Agriculture, Forest Service, 1981)

It is not clear that this extensive public involvement made the Forest Service's decision any easier. In fact, the manner in which the public was involved in the Cache Creek/Little Granite Creek process may actually have complicated the agency's task. As they acquired more and more information either about technical or ecological aspects of the proposal, or about public attitudes towards it, the agencies in many ways isolated themselves from the various interest groups. Perceiving themselves to be "the decision-maker," the two agencies set themselves apart from interests directly affected by the decision. As a result, each interest group saw its task to be one of convincing the federal agencies of the "rightness" of their position and the "wrongness" of their adversaries' positions. This approach shadowed the legitimacy of all affected groups; it provided no means for resolving the conflict that existed between them. Eventually, this approach led these groups to distrust the decision-making process and thus reject those decisions made by it. In the end, the Forest Service decided to permit exploratory drilling at both wellsites. As discussed in Chapter 4, this decision outraged several local, state and national groups and led to several administrative appeals, state hearings, and lawsuits.

The Cache Creek/Little Granite Creek case illustrates the Forest Service emphasis on public involvement in decision-making. Furthermore, the outcome of this case is representative of public dissatisfaction with this process. Although the Forest Service's approach to decision-making might give the public land managers confidence that the final assessment is in "the public's interest," it leaves the many different publics feeling unrepresented. In selecting its "preferred alternative," in the Palisades environmental assessment, the agency found, after obtaining and then analyzing public input, that this altenative

is responsive to the key issues raised by the public *since it balances the intense opposing concerns of environmental groups and the energy industry.* (U.S. Department of Agriculture, Forest Service, 1980, p. 46 [emphasis added])

The Sierra Club disagreed with this assessment and appealed the Forest Service's decision, eventually taking the agency to court.

The many publics do not perceive the USFS to be responsive to their concerns when the decisions reached run counter to their individual best interest. Although the current administrative decisionmaking process provides for *participation* by these many publics in order that the concerns of each are aired, it is not designed to accommodate their concerns *in a way that satisfies them that they have indeed been accommodated as well as possible.* Interest airing is not interest accommodation.

The Forest Service predicament persists not because it has not been acknowledged or analyzed but rather because proposed reforms have not been viable. Analyzed in the context of the long-held public-land management paradigm, past prescriptions are based on the assumption that scientifically trained land managers are capable of making these decisions, and the only question is how to bound their discretion to ensure that all interests at stake are fairly represented. As a result, most reform prescriptions as well as Forest Service procedural changes, have been inadequate. Despite extensive public-involvement efforts, despite extensive public input to and review of decisions, and despite Congressional mandates that all values be represented, frequently one user group or another is *not* satisfied that their concerns have been accommodated. These groups distrust a process that yields decisions contrary to their best interests, and therefore they oppose those decisions and appeal to other avenues for obtaining representation. To use Professor Stewart's terms, the process is a "negative" one with energies devoted to checking administrative behavior in the judicial arena rather than an "affirmative" one ensuring representation in the administrative arena (Stewart, 1985). It is a political problem lacking a political process in which the competing demands can be made and resolved through bargaining and compromise.

RETARGETING REFORM: THE PUBLIC LAND MANAGEMENT PARADIGM

Conflict is the foundation of most social problems (Coser, 1967; Himes, 1980). Bargaining and compromise to resolve disputes are the rule rather than the exception in solving such problems. If some aspects of land management have become more politicized and if power has become fairly well-balanced as suggested here and elsewhere, why do bargaining and compromise not naturally occur in the management of public lands? As the land management task developed political aspects, why did a political system not evolve to accommodate it?

Two key factors reinforce the Forest Service's traditional decision-making processes, both rooted in the strength of the land management paradigm. First, past reform proposals as well as Congressional mandates have reinforced this paradigm. Society continues to turn to scientific experts when tough problems must be solved; in this respect, the Forest Service cannot be blamed for its current dilemma. Second, the paradigm prevails because it represents the norms of the public forestry profession. Whereas Forest Service officials acknowledge public dissatisfaction of their decisions, they respond, as seen, by providing more opportunities for public input to decision-making. Decision-making itself still occurs through technical analyses consistent with this paradigm. This approach should come as no surprise considering who established the paradigm at the outset.

The culture of a profession is, by nature, self-reinforcing and stabilizing. By design, it resists external pressures for change. The profession defines its role and establishes it in society; it would be counter to the concept of professions if outsiders were able to fill this role (Mosher, 1968, Chapter 4; Kaufman, 1960). Political scientist Frederick C. Mosher (1968) explains why this is true:

> Professionalism rests upon specialized knowledge, science and rationality. There are *correct* ways of solving problems and doing things. Politics is seen as constituting negotiation, elections, votes, compromises—all carried on by subject-matter amateurs. Politics is to the professions as ambiguity is to truth, expediency to rightness, heresy to true belief. (p. 109)

Mosher has identified several common characteristics of professions. First, their purpose is to elevate or maintain their "stature and strength in the public image." Second, their objective is to expand the boundaries of work in which they have "exclusive prerogatives to operate." Third, they establish and maintain professional standards and norms through education and entrance requirements to the profession. Finally, they identify their niche by concentrating on "work substance" and expertise that sets them apart from other groups or individuals (pp. 107–131). The forestry profession is no different than other professions in this respect. That they accept responsibility for the expanded public-land management task is not surprising; nor is the way they have responded.

There has been some dramatic change in the agency in response to the demands of different interest groups for representation. Public participation programs were rapidly instituted during the early 1970s where they had previously been nonexistant. In 1970, public participation became an official Forest Service policy (Culhane, 1981, p. 232). In 1972, the agency established its Inform-&-Involve program. This program illustrated a dramatically changed perspective on the part of the Forest Service about the value of public input. It directed field staff to:

> Broaden contacts with groups, associations, and organizations to better inform and involve a wide range of the public on current programs, projects and issues.
>
> The key . . . is "awareness," and the key to awareness is "listening.". . . It means seeking out and listening to individuals and groups which may have traditionally opposed certain aspects of Forest Service management.
>
> Recognize that public involvement is an essential part of decision-making since it can enable the decision maker to render a better decision.
>
> Discard any notion that actions which will affect environmental quality or the public interest can be judged only by professionals.
>
> Keep in mind that all interest groups are champions of some aspects of good resource management. Disregarding the concern of specific groups on one issue because of extreme controversy may well weaken their desire to get involved on other issues in which they could make valuable contributions.
>
> Recognize that public involvement requires that it must be sought out *before* a decision has been reached. (Culhane, 1981, p. 241)

The Forest Service now uses what it terms "Information and Education" efforts to help groups understand different public lands issues and proposals. Public meetings and tours are held much more frequently now than in the past. One-to-one meetings with different groups are also used more often to obtain input as well as inform the public (U.S. Department of Agriculture, Forest Service, 1973). The Forest Service is hardly ignoring the problem; nor does it appear to be resisting change. The change that does occur, however, is clearly defined by professional norms and clearly within the context of scientific decision-making consistent with the land management paradigm.

As seen in past chapters, the Forest Service's public-involvement efforts have not proven sufficient to end interest group opposition to many Forest Service decisions. The agency's efforts to obtain input from "the public" do not always satisfy groups that their best interests have indeed received a fair hearing. Although official Forest Service directives now require that field staff listen to the public and keep it informed, these directives do not explain what the field staff should do with this input once they have acquired it. Is simply "considering" it all that is required? One Sierra Club director praised the Forest Service public-involvement efforts but lamented that, once decisions are made, "it is like they never listened" (Phil Hocker, personal communication, November, 1981). Whereas a problem of interest-group opposition was noted by Forest Service officials and a solution seen in public involvement, the involvement that occurred, even though rapidly instituted and thorough, has not been sufficient to avert continuing and successful opposition to decisions made.

Both Congressional mandates and past reform proposals have cast professional land managers into a comprehensive planning role in addition to their traditional scientific land-management role. There is little reason to expect that foresters trained to address conservation ends in land management will be able to suddenly adopt this comprehensive view, though. It is counter to their professional background and norms. In fact, considering these inherent professional biases, we should probably be more surprised that they have responded as dramatically as they have in providing means for obtaining public input than critical that these efforts have not been enough. Public land managers are not like comprehensive city planners who obtain information and develop alternatives that are then subjected to a political decision-making process by elected officials (Altshuler, 1965, Chapter 7). The task that has been given to Forest Service officials is to both develop these alternatives and then make the highly political decisions. That the outcome is opposition and litigation is hardly surprising. Means for making these political, as opposed to scientific, decisions have never been prescribed nor even simply suggested. Hence, the response conforms to professional norms.

CONCLUSION

The technical expert model of decision-making that has long been the hallmark of the progressive conservation ideal is not adequate when viewed in the context of today's public-land-management task. The political, resource-allocation dimensions of the decisions to be made must be recognized rather than hidden within technical analyses. In order to avoid the now pervasive and costly opposition, each interest group must be *satisfied* that its concerns have been represented. The established process is clearly unsuccessful at this task; procedural change is needed. Experts alone are not able to represent the many interests at stake, no matter how systematic, how thorough or how objective they may try to be. These disputes must be resolved through the direct involvement of affected interests. The indirect representation approach has proven itself incapable of making viable decisions; the process is no longer trusted by the many affected groups and individuals. Affected groups and individuals do not understand how specific decisions were reached and why and, furthermore, have no ownership in the decision-making process's conclusion. A reformed paradigm must not only acknowledge the limits of scientific expertise and technical analysis in decision-making but must as well include other means for making the value judgments that invariably must be made. It must recognize the legitimacy and forcefulness of the now well-distributed power and direct it towards facilitating acceptable decisions, in contrast to opposing unacceptable decisions. It must provide both the incentives and an opportunity for more cooperative and collaborative involvement of public-land user groups in agency decisionmaking.

Chapter 7 describes the objectives of a decision-making process that is purposefully structured to manage the inherent conflicts in natural-resource allocation tasks. It steps through the current planning process mandated by the National Forest Management Act of 1976 to see where and how effectively these objectives are, or might be, met. It briefly recounts several national forest planning processes that successfully accommodated the diverse concerns of specific national-forest user groups. The chapter concludes by pinpointing where procedural reforms might better address today's national-forest management task. Chapters 8 and 9 then conclude the book by proposing specific steps the Forest Service might take to begin resolving its land management disputes.

CHAPTER 7

The Opportunity of National Forest Planning

As seen in the last several chapters, conflict pervades national forest management, leading to several unfavorable consequences. Decisions are frequently not viable; they are delayed or overturned by appeals and lawsuits, leaving the decision-making process stymied. There is delay and its associated costs, particularly at the expense of on-the-ground management activities. The process is divisive; it keeps groups battling each other's demands and precludes finding a common ground should one exist. Public trust and faith in the agency and its officials continue to decline. The morale of agency staff is eroded as their efforts are constantly besieged and seldom come to fruition. There are political ramifications when external pressures are brought to bear, at times resulting in budgetary constraints, new legislative mandates, or judicially imposed direction.

Some might argue that the Forest Service is caught in a Catch-22 situation; that there is no way it can succeed given the contradictory values it is expected to serve, the militance of its constituent groups, and the character of our legal-political system. However, some change is currently occurring that raises questions about how intractable the Forest Service situation is in fact. And, this change is coming about in the context of the current national forest planning program.

THE NATIONAL FOREST PLANNING MANDATE

In 1976, the U.S. Congress passed legislation mandating a comprehensive, long-term planning effort for the national forest system (16

179

U.S.C. 1600). When he first introduced the legislation that eventually became this National Forest Management Act, the late Senator Hubert H. Humphrey commented: "To put it bluntly, we have a mess on our hands. Instead of having a comprehensive plan for the governing and protection of our resources, we have tended to focus on each problem individually" (The Wilderness Society *et al.*, 1983). He encouraged other Senators to support the bill because it was designed "to get the practice of forestry out of the courts and back to the forests" (122 *Congressional Record* S2938–45). The National Forest Management Act (NFMA) was a response to the level of conflict plaguing national forest management and a desire to have the U.S. Forest Service be more responsive to public concerns.

The regulations promulgated for implementing this act contain no fewer than ten points at which the public must be involved in agency assessment and planning. In this NFMA planning process, *regional guidelines and standards* for individual forest planning are developed in each of the nine USFS regions. Public comment is then solicited on these guidelines and standards and this input considered and incorporated as agency officials deem appropriate. Next, a *regional plan* is developed, setting the direction and targets for the individual national forest plans. Again, public comment is solicited and is considered and incorporated as agency officials deem appropriate. An *environmental impact statement* is also completed on each regional plan and is then released for public review and comment (*Federal Register, 47*, 43206).

Each national forest is now applying the direction dictated in these regional plans, guidelines, and standards to develop its individual national forest plan. Public input is solicited at the outset to help identify key issues that individuals and groups feel must be addressed in the forest plan. Agency land managers and planners then take this input and synthesize it to begin their forest planning effort. Each national forest plan, along with a draft environmental impact statement on the plan, is then released in draft form for public comment. These documents are made widely available and public review and comment encouraged. For example, in Michigan's Huron-Manistee National Forest, the early scoping effort yielded 444 different issues of concern. Agency planners then took these 444 issues and synthesized them into 25 issue categories. These 25 issue categories were then consolidated into 5 problem statements that provided the analytic framework for their proposed national forest plan and draft EIS (U.S. Department of Agriculture, Forest Service, 1985).

In practice, this planning process has been plagued by the same pathologies observed in other agency management areas. The Huron-

Manistee National Forest Plan mentioned above currently faces nine separate appeals by both environmental and industry groups. As mentioned in Chapter 1, over 600 appeals had been filed on the 75 forest plans completed as of April, 1987. The agency anticipates receiving well over one thousand appeals before concluding this *first round* of their mandated planning process. Conservation groups such as The Wilderness Society have published lengthy citizen guides on the issues to raise and steps to take in appealing national forest plans and timber sales (The Wilderness Society, 1986a,b; Atkin, 1986).

The intent of the following review of current national forest planning efforts is *not*, however, to drive the last few nails in the coffin of an agency doomed to a contentious existence. Rather, the purpose here is to better understand and to portray the planning task as the opportunity that it might be, given differing circumstances. The current contentious state is not unavoidable. Forest planning provides a tremendous opportunity for the agency to regain trust, to build effective communication with its many and varied constituents, and to develop flexible and responsive management plans. It is an opportunity to begin looking at the *issues* underlying national forest resource use rather than the resource base alone.

NATIONAL FOREST PLANNING: THE OPPORTUNITY REALIZED

For decision-making to be viable requires that national forest users be satisfied that agency actions do indeed accommodate their concerns. Accomplishing this objective, though not easy, is nonetheless possible and, moreover, plausible within the agency's own existing procedures and guidelines. As final forest plans are released and administrative appeals are filed on them, some forest supervisors are taking the initiative to redirect the course of their planning processes. Rather than steadfastly defending the plan's contents and the process used to develop them as what is professionally appropriate and in the public interest, they are instead opening the door to appellants and all other forest users to collaboratively develop a plan that *is* acceptable to all groups and does not require an agency defense. The key to their success and, moreover, the continuing good relations between all parties involved, is the cooperative process employed by the Forest Service in these few cases; it is *how* they have involved affected groups.

As of mid-1987, seven national forests in particular have success stories to share involving their planning efforts. When releasing the draft plan for public review and comment, officials on West Virginia's Monongahela National Forest immediately realized that they had misread constituent concerns. They then began an intensive, year-long collaborative process, involving representatives of various forest users in developing a mutually acceptable plan. Six other national forests—Virginia's Jefferson, New Mexico's Cibola, Alaska's Chugach, Colorado's Rio Grande, Georgia's Chattahoochee-Oconee, and the Nebraska National Forests—faced administrative appeals after releasing their final plans. Each commenced similar collaborative processes to resolve the issues under contention and, moreover, develop mutually acceptable plans. All seven plans are now finalized, have broad-based support and have not been appealed.

THE MONONGAHELA NATIONAL FOREST

One of the key factors encouraging passage of the National Forest Management Act of 1976 was a lawsuit filed by environmental groups over the U.S. Forest Service's extensive clearcutting practices in the 1960s and 1970s in West Virginia's Monongahela National Forest. This same forest's draft plan under NFMA, released in December, 1984, had timber harvest levels quadrupling and road mileage tripling. It also permitted coal mining within the forest's boundaries.

Public reaction to the draft plan was overwhelmingly negative. The comments received in almost 4,000 letters and phone calls from over 17,000 individuals came as a surprise to forest officials. Gil Churchill, Monongahela National Forest Planner and Information Officer, commented that "we expected about a quarter as much; it's the largest response we've ever gotten to anything we've done in West Virginia" (Hanson, 1987, p. 20).

The Sierra Club, Trout Unlimited, Audubon Society, West Virginia Wildlife Federation, West Virginia Highlands Conservancy, West Virginia Citizens Action Group, among other conservation and sporting groups, consolidated their efforts to educate the public about what they believed to be the plan's flaws through mailings, the media, and meetings throughout the state. According to Mary Wimmer of the West Virginia Chapter of the Sierra Club, "Our message to citizens was that communication is part of the process, and that if you participate, the Forest Service will have to listen" (Hanson, 1987, p. 21).

In response to this widespread concern, the Monongahela National Forest Supervisor at first extended the 3-month public comment period by 1 month and then decided to enlist the help of any interested groups and individuals in redrafting the plan. A year-long process ensued during which the conservation groups worked directly with Forest Service planners in redeveloping plan alternatives, gathering additional information and evaluating the alternatives. According to Churchill:

> They seriously wanted to know more—what we were facing, what latitude we had. And they were extremely positive—they gave us information on things we could actually do something about. When we got together, it wasn't just another public meeting. It was a place to do work. (Hanson, 1987, p. 21)

The new plan, approved in July, 1986, bears no resemblance to the original draft. It gives primary emphasis to wildlife and recreation in much of the forest and reduces the proposed road-building program by 25%. Timber harvest levels were cut by more than half, with those remaining receiving more stringent review and regulation than originally proposed. Public review and direct involvement is now required in reviewing coal-leasing proposals on a case-by-case basis. The plan contains a new management prescription for the 15% of the forest where roads and timber harvests are banned. In these areas the agency's primary objective will be to protect the watershed, wildlife, and dispersed recreation resource values. According to the Sierra Club's Wimmer, "This new management prescription is a highlight of the plan. It protects some of the best remaining wildlands in the Monongahela" (p. 21).

Throughout this collaborative planning effort, the Forest Service tried to no avail to bring timber industry representatives into the redrafting process. These individuals consistently refused, however. When the new plan was released, timber interests initially threatened appeals. Realizing that their complaints would likely fall on deaf ears in the agency and courts, as they had by-passed a meaningful opportunity to influence the final decisions, these timber industry representatives chose not to appeal the plan. According to Roger Sherman of Westvaco (a timber products company):

> The forestry community feels that the plan represents a backing away from commodity uses of the forest. But, with the amount of discussion that went on, there doesn't seem to be much point in an appeal. (Hanson, 1987, p. 21)

THE JEFFERSON NATIONAL FOREST

In a case quite similar to that on the Monongahela, Forest Service officials on the Jefferson National Forest in Virginia called upon the services of a professional facilitator to help them resolve several administrative appeals on their newly released forest plan. From October, 1985 to March, 1987, agency planners worked with representatives of the Citizen Task Force of Southwest Virginia, a conservation group formed in 1982 to participate in the national forest planning process, to resolve several issues in contention. Their efforts led to a formal amendment of the forest plan, highlights of which include:

- An annual conference at which the public can review and comment on activities proposed by the Forest Service
- 50% reduction in annual rate of new road construction
- Additional protection for dispersed recreation areas
- Increased law enforcement
- Expanded and clarified wildlife habitat guidelines, developed in conjunction with personnel of the Virginia Commission of Game and Inland Fisheries
- A guide to explain the planning process to the public in layman's terms
- A 50% reduction in the conversion of hardwoods to pine
- Increased research and monitoring

According to Jefferson National Forest Supervisor Thomas A. Hoots:

> Most of the changes included in the amendment respond to citizen concern for the Forest's recreation, wildlife and aesthetic values, and the desire for continued public involvement opportunities. At the same time, we have tried hard to keep in mind the needs of the other users of the National Forest. We have begun working with groups, such as the Appalachian Forest Management Group, to ensure our Forest Plan continues to have the flexibility necessary to meet their needs also. (press release, March 20, 1987)

While cutting back on planned road building, the plan kept suitable and allowable timber sale quantities the same. As a result, timber industry representatives supported rather than contested the amended plan. David Carr, representative of the Citizen Task Force, expressed the conservationist's appreciation for this type of involvement:

> This case shows what can happen when the Forest Service and conservationists work together. The settlement we have reached should ensure management of the Forest which enhances the recreation and wildlife values of the Forest. (ibid.)

Forest Supervisor Hoots concluded:

> To me it is a credit to the good faith and hard work of all sides involved that our differences could be resolved at the local level, without the need for higher level administrative review or litigation. [The] process creates a strong sense of partnership between agency personnel and all users of the Forest. These processes, by directly involving affected groups in actually gathering and analyzing data and developing and evaluating alternatives, instilled a

sense of ownership among all parties in the plan eventually adopted, a far preferrable outcome to the antagonism engendered by more traditional processes. (*ibid.*)

Similar processes have been successfully used in other national forests. In Colorado's Rio Grande National Forest, appellants and Forest Service planners agreed to changes in the monitoring program of the proposed plan. Rather than taking action on reforestation if the forest's efforts fall 25% behind established goals, they agreed on action if reforestation falls behind 10% of the plan's goals. In Georgia's Chattahoochee-Oconee National Forests, the status of one roadless area in the plan was shifted to proposed Wilderness. Three other roadless areas were given nonroaded recreation status. Timber sales were decreased 15% from 20 to 17 mmbf. Additionally, all forest-user groups will now be involved in an annual review of the plan's progress. In this case, the Forest Service did not alter their road construction targets and agreed to "minimize" but not ban the use of herbicides. In the Nebraska National Forest, the deletion of one word from the plan sufficiently addressed the concerns of appellants (O'Toole, 1986).

In yet another case mirroring that of the Jefferson National Forest, representatives of New Mexico's Cibola National Forest, the State of New Mexico, Sandoval Environmental Action Community, New Mexico Wildlife Federation, Sierra Club, Las Huertas/La Jara Ditch Association, American Indian Environmental Council, Inc., Southwest Research and Information Center, and Tonantzin Land Institute, met for 6 months to resolve their differences over an administrative appeal of that forest's plan. The final plan was first released in July, 1985, and formally amended November, 1986 (U.S. Department of Agriculture, Forest Service, 1986).

Not *all* forest planning disputes will be resolved through more collaborative, consensus-seeking processes such as those mentioned above. Many might be resolved however if only given the opportunity. These processes succeeded because they built trust between the agency and the forest-user groups participating in them. They encouraged understanding of all interests having a stake in the forest plans. They provided opportunities for joint fact-finding and collaboration, and acknowledged that the many values at stake, while different, were legitimate and therefore should be accommodated.

FIVE DISTINGUISHING CHARACTERISTICS

The successful forest planning cases mentioned above raise questions about how intractable the Forest Service's current dilemma actually is. Whereas it is certainly unlikely that there will ever be technically discernable "correct" solutions to today's forest management tradeoffs, there can be solutions (as seen in the forest planning cases described above) that will be accepted and supported by affected interests rather than contested. Value differences, of course, will always exist. However, it is sometimes possible to incorporate these different values into decision-making in such a way that affected interests are satisfied with the outcomes. Additionally, these brief cases illustrate that it is possible to rebuild trust in the agency and in its decision-making processes. It is possible as well to bolster affected groups' understanding of the process and its underpinnings, as well as to satisfy their information needs. Finally, these cases show that the decision-making process can be structured to promote cooperation and collaboration rather than adversarial battling.

Any process intended to avoid the shortcomings illustrated in past chapters, must be purposefully designed to remedy those problems promoting conflict in national forest management. Five key factors appear to distinguish the successful decision-making processes from the rest. First, the process must *build trust* in the process itself, in the agency and its staff and in the other interest groups involved. Second, it must *build understanding* of the process, of the constraints on and bounds of decision-making, of the issues involved, and of everyone's true concerns and stakes. Third, it must *incorporate the values* held by different stakeholders in such a manner that agreement can be reached on the assumptions to apply when analyzing the data and the judgments that bear on decision-making. Fourth, it must *provide opportunities for joint fact-finding* by affected groups, allowing issues and questions to be raised early and providing all parties with equal information. Finally, it must *provide incentives for cooperation and collaboration* in a problem-solving manner, rather than for continued adversarial behavior.

This process does not supplant professional land-management practices. Any decisions reached through it clearly must be professionally sound and constrained by existing legal and technical limits. However, the process is not consistent with the paradigm described in past chapters. It alters the decision-making process to provide a forum within which different forest-user groups can represent their concerns *themselves*, rather than have professional land managers attempt to do so with the shortcomings illustrated throughout this book. Such a process acknowledges that natural resource management is sometimes a political process and therefore provides a sanctioned, legitimate forum for discussion, give-and-take, and compromise. It thereby leads to a decision that all parties are a part of and, moreover, can support.

Indeed, these five factors:

• Building trust
• Building understanding
• Incorporating conflicting values
• Providing opportunities for joint fact-finding
• Encouraging cooperation and collaboration

are the objectives of *any* process whose purpose it is to manage conflict (Fisher and Ury, 1981; Moore, 1986; Raiffa, 1982; Susskind and Cruik-shank, 1987). The planning efforts of several national forests are examined below to see how well they satisfy these five objectives of a conflict-managing process, and where they might fall short of this goal. In so doing, the discussion highlights the direction that future forest-management efforts, specifically designed to manage conflicts, might take.

OBJECTIVE 1: BUILDING TRUST

It is difficult to criticize the Forest Service on its efforts to regain the trust of its many constituencies. It, probably more than most agencies, has taken to heart its public trust responsibilities and public-involvement mandates. As described in Chapter 6, it has diversified its profile to try to represent the broadening array of national forest uses. It holds frequent public meetings and hearings, and issues press releases on most decisions. Additionally, forest staff maintain personal contact with groups and individuals concerned about a broad range of forest management activities.

Although the extensiveness and, moreover, sincerity of Forest Service efforts to build trust in decision-making can hardly be denied, the effectiveness of these efforts, even agency officials acknowledge, leaves something to be desired. If the different groups and individuals with a stake in the forest planning process trusted the agency to make decisions representing their interests to the best extent possible, disputes would be minimal. Such is not the case, however. Public trust in the agency is low, following two decades of conflict. As a result, the agency is watched like a hawk and its motives and decisions are immediately questioned and criticized.

Although Bridger-Teton National Forest officials tried to keep their planning process open, the frequently revised schedule for completing the plan, combined with unanticipated reviews by Department of Agriculture officials in Washington, D.C., undermined their efforts. The plan, originally scheduled for public release in early 1982, was later pushed back to mid-1983 and then to 1985 and eventually to October, 1986. What specifically was happening with the plan during this time became more and more of a mystery to those following the process.

In December, 1985, the Secretary of Agriculture requested a copy of the not-yet-publicly-released draft plan. Douglas W. MacCleery, then deputy assistant secretary and formerly a Washington lobbyist for the Forest Products Trade Association, was scheduled to review the plan. His review was immediately labeled "hostile" by some Jackson Hole residents who felt that "it flies in the face of what is supposed to be local forest planning" (*Jackson Hole Guide*, December 10, 1985, p. 1). As Len Carlman, public lands director for the Jackson Hole Alliance for Responsible Planning (Alliance), commented:

> For better or worse, we've got to put our faith in the forest plan. The more it gets changed behind closed doors, the more it erodes public confidence. (*Jackson Hole Guide*, May 1, 1986, p. A2)

MacCleery immediately responded to these suspicions, hoping to assuage the fears of those following the Bridger-Teton planning process:

> I'm not hostile toward any interest. . . . It sounds to me as though the interests out there are pretty well polarized. In a polarized situation, it might be *doubly important that the plan be reviewed objectively to make sure everybody's concerns are addressed adequately* [emphasis added]. (*Jackson Hole Guide*, December 12, 1985, p. A12)

Conservation groups in Jackson Hole and elsewhere hardly deemed a review by MacCleery to be "objective" however. Their distrust of the process escalated and demands that the plan finally be released heightened. Again, Len Carlman summed up the community's attitude:

> This community is tired of waiting for the Bridger-Teton Forest Plan. If every *i* isn't dotted exactly in the middle and every *t* isn't crossed perfectly, we think that's less important than getting the plan on the streets. It's time to get it before the public now. (*Jackson Hole Guide*, December 12, 1985, p. A12)

Needless to say, Bridger-Teton officials were just as concerned about the decreasing public confidence and the plan's credibility as it became caught up in Washington Office reviews and changes. They argued unsuccessfully with Regional and Washington Office officials that the forest would be better off incorporating public comments and Forest Service concerns simultaneously. Ernie Nunn, acting Forest Supervisor at the time, tried to offset growing public concern by explaining that any changes occurring in the plan were merely to clarify existing meanings and make the plan more understandable to the public:

> There appears to be a lot of suspicion about what we're doing with the plan now. But what Carl [Pence, forest planner] is doing is a lot of work on presentation and readability. (*Jackson Hole Guide*, April 29, 1986, p. A6)

By the summer of 1986, the plan was being referred to as the "alleged plan" by some Jackson Hole residents. Forest Spokesman Fred Kingwill again tried to clarify the Washington, D.C. reviews and diminish fears by several groups that substantive changes might be occurring in the plan:

> They say if we're going to analyze one alternative to the nth degree, they want us to do analyses for all alternatives to the nth degree. *It's all tied to being able to withstand legal challenges* [emphasis added]. (*Jackson Hole Guide*, July 1, 1986, p. A7)

In mid-August, 1986, the Forest Service announced that adjustments to the plan were nearly complete and that the plan would be released in October. As reported in the *Jackson Hole Guide* (August 12, 1986, p. 1):

> Most of the reworking done at the behest of Agriculture Department economists and attorneys was done on Appendix B, a section of the plan which outlines the economic justifications for various alternatives. Forest Service officials have said in the past that attorneys reviewing the plan wanted the economic justifications as full and complete as possible *to help head off legal challenges* [emphasis added].

This concern about likely legal challenges to the plan had increasingly been mentioned as the reason Washington Office as well as local agency officials were trying to perfect the plan before releasing it. The often-mentioned concern about likely legal challenges did little to convince different groups and individuals that the plan would be an acceptable balancing of the many interests at stake, however. In fact, with time, the fate of the BTNF plan was assumed by all following the planning process to be decided in appeals and lawsuits. Increasingly, media reference to the plan expressed little hope for its immediate acceptance:

> But the draft of the forest plan is not expected until fall, and if it is tied up in appeals and court battles, it could take years to get the plan in place. (*Jackson Hole Guide*, July 31, 1986, p. A13)

> While Forest Service officials have tried to make the plan "litigation proof," some appeals seem unavoidable. (*Jackson Hole Guide*, August 26, 1986, p. A2)

This assumption hardly instilled faith in the plan or in the planning process, particularly one seemingly designed to avoid traveling this undesirable path.

The forest planning process is similarly plagued in other national forests. In the Santa Fe National Forest in New Mexico, public confidence in both the agency and its eventual plan was whittled away by debate over the value of additional public hearings. Several groups formally requested that the Forest Service hold an additional public hearing on the draft plan and EIS. Forest Service officials, however, refused. Ted Davis, President of the Save the Jemez interest group, argued that a hearing "would help reinstate public confidence in [the] decision-making process" (implying that public confidence had been lost) (*Albuquerque Journal North*, August 3, 1982). Jim Perry, supervisor for the 1.5-million-acre forest, argued that the request was ill-timed, coming after the deadline for public comments and, furthermore, that "additional input would not necessarily provide me anything that would make a better plan" (*The New Mexican*, September 9, 1982, p. A1). In the end, the Santa Fe National Forest Plan became deadlocked in appeals.

Forest Service actions both during the planning process (as discussed above), as well as historically in other management activities, critically influences public trust in its decisions. In Colorado's San Juan National Forest, the agency argued in their response to appeals documents that conservation groups need not worry about the environmental consequences of its proposed timber-harvesting program in the plan. In their words:

> The Forest is required by law, policy, and internal direction to produce goods and services in a sound environmental manner to meet public needs and demands. There is an abundance of laws, regulations, and policy as well as the plan itself, that contain adequate direction for assuring that adverse environmental impacts will not occur or that impacts will be acceptably mitigated. *The history of the Forest Service provides strong evidence that resources can be managed, even at fairly intense levels, without significant or long-term environmental damage* [emphasis added]. Potential environmental consequences are always considered when planning site specific projects and projects are designed to minimize adverse impacts. (Torrence, 1984, p. V-1)

However, "Forest Service history" was precisely the reason the plan's appellants were so concerned about the preferred alternative. In their appeal, they argued:

> A history of unsuccessful reforestation following past timber harvests underscores the environmental fragility of the San Juan. . . . Planners' simple assertion that the Forest Service has "modified its reforestation techniques and expect[s] substantially better regeneration" hardly constitutes convincing evidence that the serious regeneration problem on the forest has been solved. (Benfield, Ward, and Wilson, 1983, p. 42)

Appellants had lost faith in the agency to fulfill such promises and therefore had little choice but to appeal the agency's decision.

As the Forest Service has learned, the existence of or, conversely, lack of trust can have a major impact on the outcome of a decision-making process. Forest Service officials have become comfortable over the years arguing that, as they are professionals and experts, the public should trust their judgments and decisions. That trust, however, has long disappeared. Agency officials sincerely believe that they have the ingredients to develop high quality, professionally sound plans. Unfortunately, they have yet to capitalize on these capabilities by developing and following a decision-making process that encourages public trust in the agency and promotes public confidence in the final forest plans.

OBJECTIVE 2: PROMOTING UNDERSTANDING

At the heart of disputes, particularly natural resource disputes that are so complex, where so much is at stake, and where there are several legitimate yet competing interests, is often a lack of understanding. Groups do not fully understand the exact issues of concern to each party (including the Forest Service), what constraints bound the agency's considerations, and what, realistically, is the realm of possible outcomes.

Most Forest Service manuals, memos, and procedures are public information and readily and willingly made available to interested individuals. Additionally, agency decision-makers and staff, at each step of a process, outline each future step, the procedures to be followed and legislative mandates to be fulfilled. Combined with these more administrative and procedural matters, the agency holds meetings and hearings during which groups and individuals can articulate their concerns. Thus the many constituents should, in theory, be familiar with each others needs and demands.

Regardless, a true understanding is not always achieved by these efforts. Groups do not always understand each other; the agency actions behind closed doors are only procedurally, not substantively, explained by their manuals. When the substance of a decision is announced, how and why it was *specifically* derived is open to question. As seen in the forest-planning experiences described below, building and maintaining this understanding is more difficult and yet more critical than might be imagined at first glance.

The agency's actions are not meant to preclude full public understanding of the forest plans. To the contrary, their intent is to foster a broad public understanding of each forest's future. However, most attention seems to be focused on the content and presentation of the planning documents, rather than on the *process* used to develop these documents. As the BTNF plan's release date kept being extended, one BTNF official expressed what he felt was everyone's frustration:

> We're ready to let it go. We feel we've got a good document that reflects what the majority of the public wants for this forest. I'm growing more and more concerned that the longer we wait, the more credibility we lose.
>
> We're frustrated, the public is frustrated, and the only way we're going to satisfy people now is to release the thing and let everybody see it and let everybody start to deal with it. It's time. It's past time. (*Jackson Hole Guide*, April 17, 1986, pp. 1,A7)

Similarly, when the Bridger-Teton National Forest plan was reviewed by Department of Agriculture economists in Washington, D.C., their emphasis was on the content and structure of the draft plan. Their objective was to have a plan that would provide the public with the level of detail needed to understand both the plan's evolution and its proposals as well as be able to withstand legal challenges. As USDA Economic Analyst John Fedkiw put it:

> The role of this plan is to present a range of issues and address each one fully. In some cases there's a lack of information that should be included. I take no issue with their range of alternatives or their preferred alternatives. . . . I just want the most professional presentation possible with all the relevant data included.
>
> The plan assigns values to various uses of the forest, and in some cases does not provide the complete data to back up those values.
>
> The data is available from other sources, however. All we want is to have it included in the plan so that when the public needs it, everyone will understand clearly what information was used to reach certain conclusions. I'm sure the people at the Bridger-Teton understand all of the data clearly. My only interest is that they present it all in that manner for the public. (*Jackson Hole Guide*, April 22, 1986, p. A6)

This emphasis on plan content and detail, rather than on the process of plan development, however, has its shortcomings. Those with a stake in a plan's provisions do not fully understand its impact on them and therefore are not able to meaningfully participate until after the plan's release. At that time, unfortunately, much effort has already been expended by all parties, distrust has heightened and many decisions seem already to have been made.

The Forest Service admits that its approach does not necessarily succeed in adequately informing the public and that people do not always seem to understand the arguments that the agency is making. San Juan National Forest officials explained some revisions in their final plan in response to questions raised by environmental groups appealing this plan:

> This [lack of understanding regarding vegetation treatments and associated effects on timber and other resources] is not a new situation, nor is it con-

fined to just the Forest Service, which admittedly has had some difficulty articulating the need for and benefits from vegetation treatments. . . . Because of the various comments indicating an apparent general lack of understanding, the final EIS and Forest Plan were revised to more fully explain and discuss vegetation management concepts and effects which had been presented in the draft documents. The final documents also discuss the effects of failing to maintain a healthy forest on various other resources such as recreation, visual qualities, wildlife habitat, and water production. (Torrence, 1984, p. IV-19)

By focusing efforts to build public understanding on the planning documents rather than on the planning process, the agency is encouraging a paper dialogue. There is seldom an opportunity for the different groups to have meanings clarified or to directly address the questions of concern. Instead, each side picks and chooses what allegations and deletions in the planning documents they want to pinpoint.

As illustrated by these Natural Resources Defense Council (NRDC) allegations in their San Juan National Forest Plan appeals, this emphasis on the planning documents is perpetuated by the process:

The documented economic and environmental harms are compounded by the failure of the plan and EIS to disclose crucial information and by their assertion of important arguments for the first time in the final documents, depriving interested parties of the opportunity to respond. . . .

Indeed, if the analysis required by regulation were performed at all, one cannot tell much about it by reviewing either the final planning documents or the drafts which are crucial to public review and participation. . . .

. . . failure to disclose basic data before the comment period "is akin to rejecting comment altogether," something explicitly prohibited by the APA. As commentors from both the federal judiciary and Congress have noted, only if agencies "forthrightly disclose for comment" the data and policy considerations that underlie their thinking will interested persons have an opportunity to challenge those data and assumptions. (Benfield *et al.*, 1983, pp. 60, 73, 80)

Furthermore, in upholding the appellants' arguments and returning the San Juan as well as the Grand Mesa-Uncompaghre-Gunnison (GMUG) National Forest plans to their respective forests, the Secretary of Agriculture ruled in July, 1985, that:

Neither the San Juan nor the GMUG Records of Decision contain adequate explanation as to the specific nonpriced objectives or responses to public issues that will be achieved through continuing and increasing timber sales with known costs greater than expected revenues. Although non-priced objectives, such as community stability and the multiple use benefits associated with vegetation management, were discussed in general terms in the planning documents, more detailed discussion, backed by competent analysis, is

needed to inform the public why the Forest Service believes that the values of
achieving those objectives exceed the costs of the program (MacCleery, 1985,
p. 9).

This directive thereby reconfirmed the view that it is the planning *document*, not the planning *process*, that has the greatest role in informing the
public. Therefore, it demanded greater thoroughness in the analysis
contained in these documents.

It is conceivable that release of a forest plan document could be an
anticlimatic event, in some ways viewed with disinterest by affected
groups. What is important is *how* the planning effort proceeds, how
information is acquired and analyzed, how alternatives are developed
and evaluated and how tradeoffs are made. If the process is one that
directly involves affected parties as meaningful members in what is
inherently a negotiation process, then the contents of the final "agreement" should come as no surprise. The land planning and timber planning processes are inarguably complicated; the analysis procedures followed and the workings of FORPLAN (the agency's forest planning
computer linear programming model) are tremendously complex.
Groups are seldom able to understand all assumptions, constraints, and
analyses just from reviewing the planning documents. The planning
process needs to build this understanding.

OBJECTIVE 3: INCORPORATING VALUE DIFFERENCES INTO
THE PROCESS

As seen, groups and individuals value the resources at stake in
national forest planning differently, they assess impacts on these resources differently and, moreover, they weigh and balance alternative
outcomes differently. Given the same slate of resources, many different
"preferred" outcomes can be and, moreover, *are* derived by the various
affected interests.

Consequently, it is not surprising that the Forest Service's assessments and conclusions frequently come under fire. Whereas agency
officials make a concerted effort to incorporate the many different value
systems at stake, often by amalgamating them into single measures that,
they feel, fairly and appropriately balance those values, their efforts
have not met with much success.

Carl Pence, forest planner on the Bridger-Teton National Forest,
described that forest's attempt to devise compromise outcomes using
various plan "prescriptions" that seemingly accommodate the range of
concerns being expressed to the agency:

> Many interest groups have said they aren't opposed to timbering as long as it
> doesn't threaten the experiences sought by hunters and recreationists. Pre-

scription area 12 means trade-offs from the preservation side because there will be some management. But it also means trade-offs from the timber side too because development will be constrained. (*Jackson Hole Guide*, March 7, 1985, p. 1)

In response to accusations by timber-industry representatives that such proposals are not in keeping with sound, professional land management, Pence said:

It's going to mean changes in the way timber sales are laid out, but that doesn't mean it isn't professional. Obviously it's not the ideal silvicultural solution, but we're not managing the land for trees, we're managing for the needs of people and their desires. (*Jackson Hole Guide*, March 7, 1985, p. 1)

Despite these sincere efforts by Forest Service officials, different individuals still judge the resources at stake differently and therefore make different tradeoffs in their own analyses. For example, in a letter to the editor of the *Jackson Hole Guide*, the mayor of the Town of Afton wrote:

One of the responsibilities of government is to promote the general welfare of the nation's citizens. Therefore, it becomes the responsibility of the Forest Service to maximize the harvest of renewable forest resources from mature or decadent stands of timber when local economies are benefitted by such utilization. *In the process of maximizing the harvest of mature timber, the Forest Service should continue to consider other multiple uses of the forest* [emphasis added]. (January 23, 1986, p. A4)

Considering all "other multiple-uses of the forest," however, is precisely what the agency argues that it was doing in cutting back on proposed timber harvests in the first place.

Forest Service officials acknowledge the critical role different values play in their planning efforts. As seen, however, they have not yet been able to determine a way to successfully incorporate these different values into their decision-making. In late January, 1986, when release of the BTNF draft plan was believed imminent, plan details were disclosed to some conservationists and the press. The *Jackson Hole Guide* reported:

There are provisions which seem certain to provoke challenges from both conservation and development interests. In assessing the economics of logging the forest, planners used timber values averaged over several years in the late 1970s when prices were considerably higher than they are today. The use of those values is expected to be a point of heated debate when the forest plan becomes public. Oil and gas and timber interests are prepared to fight for fewer restrictions, especially in the Leidy Highlands, while conservationists and environmentalists are likely to question why the timbering scenarios were based on timber values so much higher than today's. "This is," one source said, "a forest plan that could wind up in litigation for the rest of our lives." (January 28, 1986, pp. 1, A12)

Indeed, the structure of the decision-making process invites this criticism. As each group or individual observes decisions being made and perceives their concerns to be unrepresented in those decisions, each then levels charges against the agency for failing to adequately consider their concerns.

In another situation, the Forest Service argued that its San Juan National Forest planning decisions responded appropriately to public issues because:

> once alternatives were developed to meet the overall goal of achieving a healthy, vigorous forest, an analysis was applied to determine which one *best met* this overall goal, while simultaneously protecting and managing the environment, and *responding to public issues and management concerns*. (Torrence, 1984, p. IV-34 [emphasis added])

However, the Natural Resources Defense Council appealed the agency's final decision because:

> . . . the Record of Decision, final plan and final plan EIS *have dramatically failed to satisfy appellants' concerns*, and because they violate principles of sound public policy and law. . . . (Benfield *et al.*, 1983, p. 7)

They criticized how the agency went about making its decisions:

> . . . planners failed to utilize proper procedure in allocating submarginal lands to logging for non-timber purposes and failed to produce adequate support for the particular purposes cited. In fact, an analysis of both multiple-use values and values associated with community employment indicates clearly that the harvest levels and land allocations in the plan are no more justified by non-timber considerations than by timber considerations *per se*.
> Indeed, the considerable environmental disruption which the plan will cause cannot be camouflaged by the Regional Forester's observation that, when "economic" and "social" factors are considered, the plan is "environmentally preferable" among the various alternatives. (Benfield *et al.*, 1983, p. 26,40)

The appellants further questioned the agency's choice of which values to uphold at the expense of others:

> the facts simply do not indicate that the forest's expanded timber program is necessary to avert imminent forest doom. . . . One can reasonably take the position that, over the years, nature has proven to be a pretty good forest manager, certainly the equal or better of man. One's point of view would also certainly play a major role in deciding whether a forest managed for timber production is more attractive to visitors since, while many may prefer the appearance of growing trees to dying ones, many of those would also prefer natural succession to a forest filled with a network of logging roads, trucks, skidding equipment, harvest areas and slash. (Benfield *et al.*, 1983, p. 28)

Whereas the appellants admitted that "all human activity for the purpose of forest resource extraction, no matter how protective the management practices employed, will result in some level of disturbance to the natural environment," their argument with the agency's proposal was that "the extent of disruption will be *unjustifiably* high" (emphasis added). The NRDC's own analysis obviously led to conclusions justifying quite different outcomes.

At the root of this values problem, as discussed in earlier chapters, is simply that no "right" decision exists. The wide range of different yet legitimate uses at stake leads to an even broader range of possible, yet not necessarily acceptable, outcomes. As the USFS California Regional Office noted in its planning guidelines to forests under its jurisdiction:

> There is no mathematical formula available to define the desired alternative. In fact, there are often differences of opinion about whether particular effects and outputs are positive or negative. Therefore it is important to present the results of the alternatives from different perspectives in order to provide a basis for review, judgment, and an eventual selection. The intent of this comparison of alternatives section is to facilitate a comparative understanding of the alternatives by providing quantitative and qualitative displays (narratives, tables and graphics) that differentiate between alternatives and that place the alternatives in relative perspective. (U.S. Department of Agriculture, Forest Service, 1984, pp. 1–8)

As seen, however, making these judgments explicit does not incorporate value differences into the process in a manner that is either meaningful or reassuring to those affected.

The development, evaluation, and then selection of a preferred alternative is probably the most critical phase of the planning process; in fact, it is the heart of any problem-solving effort. It is through these steps that facts are determined and analyzed and different values brought to bear on decision-making. It should therefore be the phase that most closely and directly involves affected groups as it is when their concerns and their ideas can be most effectively incorporated.

Such is not the case with forest planning, however. Alternatives are not creatively molded to address the specific issues and concerns surrounding a particular situation, but are rather designed to merely represent a spectrum of potential outcomes. In so doing, the process seems, so to speak, to be forcing the foot to fit the shoe rather than making the shoe to fit the foot.

The California Regional Guidelines contain instructions for developing a *range* and *types* of alternatives in that region's environmental impact statements and forest plans. The following is the direction provided for developing alternatives that deal with wild and scenic river considerations (U.S. Department of Agriculture, 1984):

> Include and evaluate each eligible river for its suitability for designation in a wide range of alternatives. . . .

1. Different alternatives should present a varied range of Wild and Scenic River recommendations that appropriately reflects the theme of each alternative.
2. In at least one alternative, recommend designation of all eligible segments of the river at their highest classification.
3. In two other alternatives, recommend two different combinations of eligible segment designation less than the entire river (e.g., segs 1, 2, 3 in one alternative; segments 2, 3 and 5 in another alternative).
4. In one alternative, if appropriate, change the classification from the highest eligibility to a lesser classification to accommodate future planned projects (e.g., a proposed road crossing would be permissible in a scenic or recreation, but not a wild, classification).
5. In two other alternatives, show at least two different management strategies for the river corridor without wild and scenic river designation. (Pp. 2-12–2-13)

At no point does the direction suggest that the key issues of concern with a particular river be discussed along with how these specific issues might then be directly addressed. At no point does the agency's own guidance encourage creative solutions, designed specifically to fit with the particular problems posed by each management situation.

The process of allocating scarce resources involves politics and value judgments, that is, trade-offs between different legitimate yet competing interests. As structured, however, the forest planning process does not provide an opportunity for affected interests to systematically ensure that their concerns and their values are integrated into decision-making. Instead, they must take the agency's determinations on faith; and, as seen earlier in this chapter, this faith is not strong.

Because so much is at stake, this process of incorporating the many different values involved into the forest planning and management process is a critical one. A more direct process than that currently used by the agency is needed. It must be an open, concerted process that satisfies groups that *their* concerns are part of the accounting. It is a process that must directly involve the different groups in incrementally developing, mutually agreed-upon measures and criteria that will be used in making the tradeoffs and balancing the resources affected by decision-making. To do otherwise will only perpetuate the debate and the divisive battles described earlier.

OBJECTIVE 4: PROVIDING OPPORTUNITIES FOR JOINT FACT-FINDING

To facilitate both meaningful and satisfying participation by national forest users in agency decision-making requires that each group and individual be operating with equal information. If different ideas are being raised, based on differing assumptions, disputes will likely persist. Furthermore, the longer a group operates under differing assumptions and with varying information, the greater the likelihood that their beliefs and positions will become cemented in place, providing little room for movement in later discussions.

Merely being informed that certain levels of resource outputs are "preferred" does not convince those with a stake in the planning to accept those decisions. Not only do they need to understand the implications of different outputs, but they need, as well, to be a part of the process that goes about obtaining and analyzing this information. For example, the Forest Service announcement in January, 1986, that 15.9 mmbf would be the draft forest plan's proposed timber-harvest level led to several predictable responses. These responses underline the importance of joint fact-finding and the likely difficulties should the process not provide for it. Darrel Hoffman, Board Member of the Jackson Hole Alliance for Responsible Planning, commented that:

> The importance of the number is whether it protects the nontimber values on the forest. If that's 15.9 mmbf, so be it. We think the key values on this forest are wildlife and scenic values, and we are prepared to fight for those values if we have to. (p. A2)

The Sierra Club's Phil Hocker responded similarly:

> The importance of the number is whether it protects the nontimber values on the forest. If that's 15.9 mmbf, so be it. We think the key values on this forest are wildlife and scenic values, and we are prepared to fight for those values if we have to. (p. A2)

Jim Hendrickson, general manager of the Long Tree Timber mill in Afton, Wyoming, threatened that the mill would be shut down if the BTNF forest plan failed to accommodate their timber needs:

> We just want a feasible, long-term cutting program on the Bridger-Teton. We don't want to hurt wildlife or scenery or tourism. We enjoy all those things, too. We want a realistic, viable timber program, and if we don't get it, we'll close this mill down. (*Jackson Hole Guide*, March 13, 1986, p. A11)

And, Jim Robinson, one of the founders of Dubois' Citizens for Multiple Use, concluded:

> It's a good thing that there's plenty of time for public input because it's clear to me they need more. (*Jackson Hole Guide*, January 30, 1986, p. A2)

Later, when the draft version of the Bridger-Teton National Forest Plan was released in October, 1986, John Barlow, President of the Wyoming Outdoor Council, felt compelled to ask the same question, a question that the plan was to have answered yet did not:

When they asked the computer how to generate the most money from all uses on this forest, it said, "Cut 2.2 mmbf." When they asked it just to maximize timber income at sustained levels, it said, "Cut 3.1 mmbf." And yet they call for 15.9 mmbf. How they came up with this number is described nowhere in the plan. We're forced to assume that it was generated entirely by political considerations. (*Jackson Hole Guide*, November 13, 1986, p. A3)

The Forest Service consumed considerable time, energy, and money making their tradeoffs; it took 6 years of work and cost $2 million to complete the draft plan. Once again, however, the process is not structured in such a way that the different groups will be convinced that the proposed timber output is "realistic" and "viable," while protecting "nontimber values." There is never an opportunity for the different groups to jointly determine where and how timber resources are available and with what consequences. The how and why of the final decisions is not clear to the groups directly affected by these decisions. As a result, opposition becomes inevitable.

The situation is the same in the San Juan National Forest case. Plan appeals there questioned the data the Forest Service brought to bear on its decision-making. Appellants argued that agency considerations were both inadequate and incorrect:

Planners' characterization of road- and timber-related soil erosion that exceeds tolerable limits as a "short-term effect" also deserves rebuttal. Such excessive erosion is a short-term *process* perhaps, but the effects on productivity may be indefinite.

The plan has failed to identify and exclude from the forest's timber production base those lands which are unsuited for timber production when economic factors are considered. Instead, the plan has designated as suitable for, and allocated to, timber production *much land that is essentially worthless* [emphasis added] for that purpose. (Benfield *et al.*, 1983, p. 55)

The San Juan National Forest Plan appellants called upon several well-respected foresters and forest economists who corroborated their findings:

The benefits are not only overstated, but many may be entirely illusory . . . much of the land on this forest is not economically suitable for timber production, and certainly cannot support an expanded timber program. (Arnold Bolle, Dean and Professor Emeritus of the University of Montana School of Forestry, pp. 36–37)

Community employment considerations do not come close to justifying submarginal sales from the San Juan. (William F. Hyde, Associate Professor of Forest Economics at Duke University's School of Forestry, p. 37)

In concluding their appeal, the appellants summarized their belief that the Forest Service's decisions in this case clearly diverged from what the facts of the case would have suggested:

> No matter how the San Juan data are analyzed, the conclusion is inescapable that lands which are economically unsuited for timber production have been allocated to that use. This is clear from the planning documents themselves as well as from expert analysis. The allocation of unsuitable lands occurs in spite of a clear history of past economic losses from the forest's timber program and in spite of the evidence indicating that, even when nontimber values are considered, the plan's program cannot be justified. The plan simply does not make good economic sense. (p. 39)

The Forest Service responded predictably that their decisions were indeed based on the facts at hand and that the appellants did not truly understand the situation:

> In all cases, impacts are demonstrated to be not as significant as Dr. Bolle suggests. They are amply discussed in the final EIS, and capable of being dealt with through appropriate preventive and mitigation measures.
>
> As noted in Chapter V of this Responsive Statement the appellants relied on general assertions and applied them to the Forest. They failed to acknowledge or consider that the Forest Service developed specific mitigation measures to bring impacts within legal and policy constraint limits and disclosed the consequences of implementation of all alternatives in the final EIS, Chapter IV. Mitigation measures are the management requirements (general direction and standards and guidelines) displayed in Forest Direction and management area prescriptions in the Plan on Pages III-11 thru III-291. The management requirements are designed to protect Forest resources and mitigate adverse impacts; therefore, the alleged "substantial environmental harm" simply will not take place [emphasis added]. (Torrence, 1984, pp. V-35, VI-2–3)

At one point in the debate, NRDC argued that "the overall favorable present net value of the plan may be due to inclusion of a number of questionable assumptions in the economic analysis" (p. IV-10). The USFS reply to this criticism was that "Although they imply that questionable 'assumptions' were used, they did not submit other assumptions that might be better" (p. IV-11). At no point does the process internalize this dispute, however. It never encourages analysis and adoption of mutually acceptable assumptions under which analysis will then take place. It never provides a forum within which the facts can be jointly determined and analyzed. Consequently, the debate is shifted to the review and appeals process, only promoting further disagreement and misunderstanding.

Jointly determining what information is needed and then *jointly* acquiring and analyzing this information, has several advantages. It forces the groups to come to grips with and articulate their true issues of concern and those assumptions on which they are basing their decisions. It forces them to listen to the concerns of others, to test each others assumptions, and to have a legitimate forum within which to make adjustments in their arguments to accommodate each others' needs. It allows them to understand and account for the legitimate concerns and needs of other groups. And, it places everyone on an equal footing, understanding the full resources at stake, the ramifications of different decisions and the constraints bounding the realm of possible outcomes. Moreover, it promotes each group's ability to contribute meaningfully to the process, make creative suggestions, articulate their different assumptions and jointly develop a mutually satisfactory outcome when possible. While it may be more difficult logistically and procedurally to do so in the short run, the long run benefits, as seen in those planning efforts described earlier in this chapter, should more than compensate for the early difficulties.

OBJECTIVE 5: PROVIDING INCENTIVES FOR COOPERATION AND COLLABORATION

It is the structure of the decision-making process, and, moreover, the incentives it places before the various interest groups, that determines how affected groups will participate and whether or not their efforts will be constructive ones.

The Forest Service's traditional decision-making process has the agency trying alone to achieve a balance between many competing interests. As a result, the process encourages a negative, adversarial behavior on the part of each interest as each seeks to convince the agency decision-makers of the validity and legitimacy of their concerns, as opposed to those of their adversaries. It seldom provides an opportunity to set adversarial behavior aside and to focus on the key *issues* of concern rather than those *positions* that might best influence agency decision-making (Fisher and Ury, 1981; Raiffa, 1982; Susskind and Cruikshank, 1987). It provides few opportunities for collaboration between competing groups.

Because the process is structured only for input and then criticism and *not* for constructive development of ideas and solutions to problems, people do not have to grapple with the very real budget, labor and resource constraints that confront the Forest Service or with the very real concerns and interests of other land users. Therefore, there is no incentive to develop demands within this framework; in fact, the incentive is just the reverse. As it is the Forest Service that has to make the decision and the allocations, each group has the incentive to argue for as much as possible in hopes of maximizing what, in the end, they receive.

At a meeting held in the Town of Afton, acting Bridger-Teton National Forest Supervisor Ernie Nunn and Local District Ranger Jerry Hawkes assured residents that the level of timber harvests proposed in the yet-to-be released forest plan would be more than necessary to sustain their mills. However, *specifics* were not and could not be discussed prerelease of the draft plan. Jerry Hawkes explained:

> I can't tell you what the exact figure [of mmbf timber to be harvested] in the forest plan is, but it's pretty darn close to the average of the last eight years. Maybe just slightly below. (*Jackson Hole Guide*, April 8, 1986, p. A3)

Despite these assurances, Afton residents did not want to take any chances. They were observing the efforts of groups in other communities to influence the outcome of the forest plan and felt that they would be foolish to be complacent now, merely accepting the agency's promises. They only had general notions of what, precisely, the Forest Service would be proposing. With environmentalists and other groups continuing their lobbying efforts, some Afton residents felt they should continue doing so as well. One local outfitter encouraged his friends and neighbors to actively fight for the timber resources needed to keep the Afton mills operating:

> If we're going to have an impact, we've got to keep on top of this thing because the environmentalists are going to be in the Forest Service offices every day with their plans. We can't let that happen. (*Jackson Hole Guide*, April 8, 1986, p. A14)

The process seems to breed adversarial behavior. It provides no forum for these groups to more proactively and cooperatively raise and address their concerns; instead, groups must react to Forest Service actions and do so in a combative manner.

When Reid Jackson retired as Bridger-Teton National Forest Supervisor in late 1985, he expressed the following hopes for the future of the forest's plan:

> I'd like to see some of the commodity uses accepted, and an acceptance of the timbering program together with the recreational, wildlife, and scenic uses. If we work at the plan and at explaining it, and if people are willing to compromise, we can get it accepted without a lot of legal appeals.
>
> But right now there is a lot of polarization out there. Big Piney has gone for commodity use, and the people seem happy with that. Folks down in Sublette County don't want commodity usage. They're very protective of what they have just as it is. The interests in Fremont County are in timbering, and the sense in Jackson Hole is preservation.

> We can use this forest plan to bring people together if only everyone will
> compromise a little bit. It's complex, and we're going to have to work real
> hard to explain it and get it accepted. (*Jackson Hole Guide*, November 19, 1985,
> p. A14)

Unfortunately, no forum has ever been provided to end the polarization
and encourage the compromises Jackson recommends. In fact, whereas
the hopes of Bridger-Teton National Forest officials mirror those expressed by Jackson, the process works against them.

In December, 1985, the Forest Service released an inhouse study by
three planning and economics specialists analyzing the implications of
management changes on the Bridger-Teton National Forest. It depicted
the forest planning process as a win–lose proposition:

> Scenarios developed for the forest plan propose either development with
> associated timber sales, or preservation for the Upper Green and Mt. Leidy
> Highlands. User groups from one or more of the major areas will be threat-
> ened regardless of which scenario is implemented.
> The dilemma is apparent. Either the L-P mill at Dubois is sustained, with
> coincident losses to user groups tied to the other resource uses on the forest,
> or the mill will not be sustained, with coincident losses in jobs in both Dubois
> and Riverton. As with most difficult social choices, some people will win
> while other people will lose. (*Jackson Hole Guide*, December 10, 1985, p. A3)

To the many different groups following the Bridger-Teton planning process, the report seemed to reconfirm the need for continued position-
taking and adversarial battling. Fred Kingwill, forest spokesman, immediately disagreed with the report's conclusions:

> It depends on how you define winning and how you control the losses. I'm
> still at the stage of feeling that some of the stuff can be mitigated. The losses
> and the impacts can be less. There will always be some losses, but they can
> be reduced. (*Jackson Hole Guide*, December 10, 1985, p. A3)

Although Kingwill's suggestion was meant to ease the growing tensions, the planning process itself did little to accommodate his notions.
Although finding a common ground and employing mitigation measures are the keys to offsetting some of the impacts of the planning
changes, the process did little to encourage such efforts.

This is not to say that Bridger-Teton National Forest officials have
not tried to establish a constructive dialogue with all interested groups
and individuals. According to the *Jackson Hole Guide*, "The Forest Service
has allowed groups and individuals to review preliminary work on the
plan, hoping to avoid problems and delays when the draft is released"
(August 29, 1985, p. 1). Randal O'Toole, a forest economist with the
nonprofit forestry-consultant organization, Cascade Holistic Economic
Consultants, has critically reviewed several forest plans, and was called
upon by the Jackson Hole Alliance to review a not-yet-released draft of

the Bridger-Teton plan. The Forest Service was receptive to the review and permitted O'Toole access to the planning documents. In fact, the forest's planner, Carl Pence, expressed appreciation for the review:

> Although there were a few items in the plan which made him uncomfortable, it was a positive experience for us. Story Clark [of the Jackson Hole Alliance] and I plan to meet in the near future to discuss O'Toole's formal recommendations. After that we'll have a better idea of what to do. (*Jackson Hole Guide*, June 21, 1984, p. A11)

Such efforts are far beyond how things have traditionally been done in the Forest Service; they are the first steps towards truly reforming the forest planning process and addressing the problem it poses. Unfortunately, these intermittent efforts cannot alone remedy the situation. More formal and concerted processes are needed.

In mid-1985, some Dubois residents formed the "Citizens for Multiple Use" group to fight the forest plan should it adversely affect their livelihood. Dubois is home to a Louisiana-Pacific mill that will likely be shut down by a forest plan calling for reduced timber harvests. Dr. Jim Robinson, superintendent of schools in Dubois and a founder of Citizens for Multiple Use, described the group's concerns:

> We would like to see a forest plan that can accommodate both sides, however that can be done. We don't expect the environmentalists to pack up and leave, and we would hope the timber company doesn't pack up and leave. But the forest plan needs to allow us to survive without severe economic impact.
>
> If we're given the time to adjust to changes we know are coming in the next five to six years, we probably could do it, but if the big changes are going to fall on us in the first three, four or five months of the forest plan, we can't handle that. People here will just pack up and leave. (*Jackson Hole Guide*, October 15, 1985, p. A3)

The Sierra Club's Phil Hocker responded to Jim Robinson's suggestion that accommodation of all interests may be possible:

> The Forest Service always has leaned over backwards on Louisiana Pacific's side of things. That mill has been overcutting in a progressive radius around their home base for a long time. If their survival means overcutting on a continued scale, then I don't think there's a way to insure their survival. *If they can survive with less cutting, then we'd be interested in talking to them.* (*Jackson Hole Guide*, October 10, 1985, p. A3 [emphasis added])

Robinson describes his group's future plans should an alternative dialogue not succeed:

it would be a matter of community education, letter writing, phone calling, public forums, direct contact with the decision-makers. If there are more formalized routes we can take, like legal appeals, we will take them, too.

We're not representatives of the timber industry. I don't think this group would favor massive forest cutting expansion, either. We don't want to see the forest so debauched that we can't hunt, fish, camp, snowmobile and ski. We all do that. None of us would live here if we didn't do those things. What we have to do is strike a balance . . . both sides have gone to an extreme right now and that's the basis of our problem. (*Jackson Hole Guide*, October 15, 1985, p. A8)

He later lamented:

I don't mean to be trite, but unless we do some mitigating, we are not going to survive. (*Jackson Hole Guide*, January 21, 1986, p. 1)

Others echoed Robinson's desire that mitigation measures be explored. Dubois Mayor Danny Grubb commented:

We don't believe we have to have a win/lose scenario. We believe there's a win/win scenario out there somewhere. We don't know what it is yet, but we're looking for it. (*Jackson Hole Guide*, January 21, 1986, p. A11)

Similarly, Stuart Thompson, a director of the Wyoming Outdoor Council, said:

I tend to agree with the optimists over there who say the forest plan doesn't have to be a situation in which some communities win and some communities lose. If we all pull together and this doesn't deteriorate into a Pinedale-versus-Dubois or a Jackson-versus-Dubois thing, we can work it out. But somebody is going to have to change the basic way of doing business. (*Jackson Hole Guide*, January 28, 1986, p. A11)

At this point, however, the process has not provided an opportunity for this effort to meaningfully occur; "the basic way of doing business" remains the same.

In April, 1986, acting forest supervisor Ernie Nunn announced that a timber conference was planned that would involve *all* users of the Bridger-Teton National Forest who wanted to attend:

It would include all of the mills in northwest Wyoming, plus county and community representatives, outfitters, guides, recreational interests, ranchers and anyone else who is interested. (*Jackson Hole Guide*, April 8, 1986, p. A14)

The meeting would be held after the draft plan was released and its objective, as reported in the *Jackson Hole Guide* (April 8, 1986, p. A14) was to have "the diverse interests on the forest to get together and try to hammer out forest plan compromises in an effort to keep the plan out of the appeals process and possible litigation." Nunn said he was prepared to use professional mediators to try to resolve the differences:

> These people are our stockholders. We owe each individual as much as we owe any special interest group. If anyone has a question, a problem or an opinion, he expects a response and he will get it. That's a promise. (*Jackson Hole Guide*, March 13, 1986, p. A15)

> We don't want to talk about selling the plan as though we absolutely know that the draft, as written, is the best thing for the forest and everybody who uses it.
>
> We don't want to use language that makes it sound like we'll be trying to shove something down the public's throats. We think this plan is the best effort we could make to reflect what the majority of the public wants for this forest, but if we're shown that it isn't, we'll change it.
>
> It's a document that's probably not going to satisfy all the needs of the local and national public. We are going to have to tell some folks some things they aren't going to like to hear, and maybe we won't be able to resolve those conflicts alone.
>
> We're managers, and sometimes we don't operate well in social situations, handling social problems. If we have trouble, we will have to find outsiders who can help us, either Forest Service or outside consultants who are experts in conflict resolution. It's something we're fully prepared to do if we need to. (*Jackson Hole Guide*, March 13, 1986, p. 1)

Nunn's announcement acknowledges the difficulties in, yet importance of, having the plan be supported. Although coming somewhat late in the planning process (after distrust has escalated and adversarial relationships are well-established), the announcement nonetheless indicates a growing interest in more direct involvement of affected groups. Unfortunately, as of mid-1987, such a meeting had yet to be held.

When finally released for public review and comment, the Draft Bridger-Teton National Forest Plan drew mixed responses. The timber industry was predictably upset with its provisions. According to Bob Baker, Louisiana Pacific's Chief Forester in Dubois:

> Based on what we've heard so far, the plan is far too extreme for us to accept. I expect a major fight over the plan.
>
> The plan would reduce the total timber harvest, close the Union Pass Road to us and move what's left of the timbering from the north to the south end of the forest, pretty much out of our reach. That seems pretty extreme to me. I would certainly be surprised if it doesn't go all the way to the courts. (*Jackson Hole Guide*, October 7, 1986, p. 1)

Environmental groups, however, expressed relief at its contents. John Barlow, President of the Wyoming Outdoor Council, commented:

> We're impressed, but not surprised. The Forest Service is doing what it said it would do. . . . . in terms of gross elements of the plan as released in the

draft, it's not inconceivable that we could come up with something we wouldn't even have to appeal. (*Jackson Hole Guide*, October 7, 1986, p. A11)

Their major concern at that point was that apathy would allow timber and other commercial-use pressures to cause unfavorable changes between the draft and final versions of the plan. As Barlow put it:

> It's a lot easier to get people to write letters opposing something than supporting it. If we don't generate enough letters to the Forest Service to support the preferred alternatives, the other side will and the Forest Service does weigh its mail.
>
> Our real enemy is complacency. I'm scared to death that people will declare victory, dust off their hands and go home. (*Jackson Hole Guide*, October 7, 1986, A11)

Similarly, Phil Hocker urged action on the part of conservationists:

> We have to read this plan very carefully. We've got to guard against a situation in which Bob Baker gets everything he wants, and the elk get the rest.
>
> We should expect that L-P would be in there trying to swamp the plan to turn it back to their liking, and we're just going to have to outshoot them. I think it's critical that everybody understand that.
>
> People who've come out well will be complacent, and people who have come out badly will come off the wall, and they could win. We're clearly vulnerable to that sequence of events. (*Jackson Hole Guide*, October 7, 1986, A11)

Three weeks later the environmentalists adopted a predictable stance. They called the plan "a sad hoax" and began criticizing its provisions. Phil Hocker issued a statement in which he said:

> Great lip service has been paid to wildlife and recreation in the Forest Service explanations of the plan, but the reality as defined by the plan will be business as usual—continued prostration of the wild heritage of western Wyoming to a dying timber industry and moribund oil and gas exploration. (*Jackson Hole Guide*, October 28, 1986, p. 1)

The Sierra Club and Wyoming Outdoor Council expressed distrust in the plan's "ambiguous" language and questioned the Forest Service's sincerity in reducing future timber sales. A *Jackson Hole Guide* editorial encouraged activism on the part of Jackson Hole residents:

> It is not an understatement to say that nothing—absolutely nothing—is likely to have more long-range effect on the economic welfare of the citizens of Jackson Hole than this forest management plan.
>
> The Forest Service is going to be influenced a great deal in drawing up its final plan by its perception of what the *public* wants. This is our chance to tell the Service, in no uncertain terms, what Jackson Hole wants and needs.
>
> After Jackson, there will be other meetings, including one in Dubois and one in Afton, where timbering interests will make their pitch.

> We can't afford to let Forest Service personnel go into those meetings
> without knowing where we stand, lest they come out of them believing that
> timbering should play a larger role than envisioned in the management plan
> draft. (November 11, 1986, p. A4)

And, the battle still wages. The process promotes such a response if the groups are going to ensure that their interests will truly be accommodated by the plan. The incentives provided are for this continued adversarial battling, not for complacent acceptance of agency recommendations.

To transform the process from an adversarial one that encourages battling through appeals and lawsuits to a potentially more collaborative one, the incentives provided by the process need to be changed. The process should provide incentives to affected parties to cooperate and collaborate in jointly solving the problem at hand rather than battling for outcomes that are win–lose and solve the problem only suboptimally. It should then provide a forum wherein meaningful dialogue can occur, compromises can be reached and affected groups can acquire some ownership in the final decisions reached.

CONCLUSION

Forest planning inarguably poses a critical and incredibly complex problem to which there are no technically or silviculturally "correct" solutions. There are, nonetheless, a multitude of potential solutions that do address all needs to varying degrees. Many of these potential solutions are never developed or considered, though, because there is no forum that encourages or accommodates such an effort. At no point does the forest planning process acknowledge that the problems to be addressed are *mutual* problems, shared by the Forest Service and all groups with a stake in national forest management. As a result, at no point does the process provide for mutual efforts towards developing solutions to these problems.

As seen in this chapter, with just a handful of exceptions, the forest planning process fails quite dramatically to satisfy those objectives determined earlier to hold some hope for its success. Whereas the agency outwardly tries to build trust, cooperation, and faith, the process used undermines their hopes by eroding all three. Whereas the planners promise opportunities to reach consensus, the process provides no forum. Whereas the individuals and groups involved keep raising what they feel are the underlying issues that need to be grappled with in developing the plans, the process encourages them to adopt positions and pursue adversarial avenues in hopes of, indirectly, satisfying their concerns. Whereas the planners make statements about the carrying capacity of the forest for timber and other commodities given the values of the other resources at stake, the process does not divulge specifically how such data were acquired and conclusions reached. Whereas some individuals try to initiate dialogues with other groups and the Forest Service, these dialogues are seldom officially sanctioned and therefore seldom prove viable or effective.

The scope of this National Forest Management Act planning effort, the number of times the public may provide input, and the issues being addressed, are all markedly different than the earlier agency planning processes. Unfortunately, *how* these tasks are accomplished is the same, and therefore the process fails for the same reasons. The process' demise is rooted in the overriding attention given to the final planning document, rather than to the *process* of planning. Consequently, the opportunity that the forest planning mandate presents for a more harmonious and effective forest management strategy has not been seized upon by the agency.

Alternative forums *can* serve to resolve differences in a manner acceptable to all parties, including the Forest Service. Doing so at this point, though, requires supplementing more traditional review and analysis procedures with more direct, collaborative efforts involving concerned forest users. Chapter 8 recommends specific reforms to the Forest Service decision-making *process*, reforms designed to manage the conflicts inherent in today's national forest management task. These reforms build upon the five objectives presented in this chapter—to build trust, encourage a broad understanding, incorporate value differences, provide opportunities for joint fact-finding, and encourage collaboration and cooperation. Chapter 8 draws from a growing theory and practice of environmental conflict management to illustrate how other, similar disputes have been resolved. The chapter illustrates the successful application of these concepts to three specific forest management disputes.

CHAPTER 8

The Importance of Process in Resolving National Forest Management Disputes

Since the mid-1960s, turmoil in national forest management has been, for the most part, unavoidable. It has been the product of changing societal demands for both commodity outputs and recreation and scenic amenities from the national forests. It has been the product of a dramatically advanced technology and, moreover, a greater awareness, understanding, and appreciation of the wilderness, wildlife, and other natural values harbored by the national forests. And, moreover, it has been the product of a cavalcade of new national and state legislation altering the mandates before the U.S. Forest Service.

Such dramatic changes, all occurring in a relatively short period of time, were bound to set the existing system spinning. They were bound to force a reassessment of management practices and the processes used to make decisions. Furthermore, the clash of different interests was unavoidable as each sought to both protect what they had already invested in the national forests and to gain what they now felt was rightfully, and legally, "theirs."

Today, however, a new equilibrium is being set. The many different interest groups have made their positions and concerns known and have achieved important precedents through both administrative and judicial rulings. The agency is now faced with implementing its new mandates under this new order. Insights into how it might begin doing so can be found in processes currently being used to resolve disputes in

211

some current national forest management efforts as well as in other environmental arenas.

ENVIRONMENTAL DISPUTE RESOLUTION

The idea that some environmental disputes might be managed and resolved is not a new one. Since the mid-1970s, academics and practitioners have explored ways in which these conflicts might be resolved through cooperative, consensus-building processes rather than in the traditional, adversarial manner that current processes often encourage (Bacow and Wheeler, 1984; Bingham, 1986; Crowfoot, 1980; Harter, 1982; Mernitz, 1980; Susskind and Cruikshank, 1987; Talbot, 1983). Disillusionment with the results of court decisions along with the limits of administrative appeals have led some traditional adversaries to the bargaining table in hopes of achieving a more desirable outcome (Susskind and Weinstein, 1981). Moreover, the costs involved in current appeals procedures and judicial reviews are encouraging environmental and community groups, development and business interests, and public agencies to seek other means for resolving their differences. These alternative processes are leading to more creative, problem-solving sessions and outcomes often not possible in the traditional administrative or judicial proceedings. And, in turn, past successes are attracting other groups in additional cases to try resolving their differences rather than automatically proceeding to the courts or political forums.

Many national forest management disputes fit the description of disputes amenable to resolution. The different parties having a stake in the decisions to be made are well known and organized. The issues of dispute are well-defined. Power between these parties has become well-developed and balanced through lawsuits, administrative rulings, and Congressional mandates. These management and planning decisions inevitably must be made. The different parties to the disputes have exhausted other avenues by which to obtain representation to their satisfaction. It is costly to all parties to continue in an adversarial process, never focusing on or resolving the real issues of concern.

The process I am proposing (see pp. 228–241) is *not* designed to resolve the fundamental value differences, like those separating groups such as the Sierra Club, Wyoming Outdoor Council, or The Wilderness Society from groups such as the Mountain States Legal Foundation, Rocky Mountain Oil and Gas Association, or Louisiana-Pacific Corporation. The process also will not create universal agreement on the proper allocation of public land resources. These value differences will always exist. Gerald Cormick (1982) of the Institute for Environmental Mediation in Seattle, Washington makes the distinction between conflicts and disputes. He defines *conflicts* as occurring "when there is a disagreement over values or scarce resources", and a dispute "is an encounter involving a specific issue over which the conflict in values is joined." It follows, then, that "resolution of a conflict is achieved when the basic value differences that separate the parties are removed," and "settlement of a dispute is achieved when the parties find a mutually acceptable basis for disposing of the issues in which they are in disagreement, despite their continuing differences over basic values" (Cormick, 1982, p. 3). To be viable, decision-making must acknowledge the legitimacy of each set of concerns. It must assure each group that their best interests are represented in the decision made. The process proposed here is designed to settle the national forest management disputes that arise because of these underlying value differences.

Environmental dispute resolution is not merely the subject of academic thought and discussion. The examples of successful attempts at all levels of government to resolve environmental disputes are numerous. (Bingham, 1986) Such a list, although too long to detail here, would include cases involving wilderness and proposed wilderness areas and endangered species, cases in which, at first glance, all parties involved would confidently proclaim "There is no middle ground!" The list would even include an oil and gas exploration proposal in a national forest under consideration for wilderness designation. The reason that such disputes over so-called nonnegotiables can be settled is *not* that one side "sells out" or "caves in" to another. Rather, by substituting a cooperative, consensus-building process for the more traditional, adversarial process, groups can focus on issues of direct concern to them rather than on positions with which they might "win."

For example, when the endangered Whooping Crane was threatened by construction of the proposed Grayrocks Dam in Wyoming (Wondolleck, 1979), or when the endangered Sandhill Crane was threatened by construction of Interstate 10 in Mississippi (Golten, 1980), predictable battle lines were drawn. Environmental interests fought to stop both proposals in order to protect the endangered species. Development interests sought to obtain approval for the proposals by arguing for the benefits to be achieved. Although costly and time-consuming court battles were forecast for both disputes, the eventual outcome differed markedly. Dispute resolution processes were used to change this position-orientation of the disputants (disapprove the dam/approve the dam) to an issue-orientation (how can the endangered species be protected and the energy or transportation needs be satisfied?). By changing their orientation, the disputants were able to then focus attention on ways in which all concerns could be addressed. This approach broadened the agenda of alternative solutions and eventually led to settlements that would not have been achieved in the courts. Similar processes have been used to successfully resolve disputes over the allocation of resources in Idaho's Gospel-Hump Wilderness Area (*Congressional Record*, October 20, 1977, S17375–17400), to develop a consensus on Colorado's wilderness designation recommendations, to protect an endangered bear in Alaska'a Kodiak Wildlife Refuge (*National Wildlife*, 20(1), 1982, p. 32G), and to permit oil and gas exploration in a Wyoming Further Planning Area while still preserving the area's wilderness characteristics.

To illustrate the progression of events in a process designed specifically to manage conflict, three such processes representing varying levels of formality, are described below. The first case involves a dispute over a proposed road and associated timber harvests in Colorado's San Juan National Forest. The second case occurred in Oregon's Willamette National Forest and involved that forest's 5-year timber management plan. The third case involved an exploratory oil and gas well in a further planning area in Wyoming's Bridger-Teton National Forest.

THE SAN JUAN NATIONAL FOREST MEDIATION

In early 1981, the USFS announced its plans to construct 50 miles of roads in a 23,000 acre section of Colorado's San Juan National Forest to support timber harvesting. Prior to making this proposal, agency planners completed an analysis of the area's resources and the impact of the road and subsequent timber harvesting on this resource base. They surveyed and marked the proposed road location. They then asked for public comment on their proposal.

The area, as would be expected of a national forest in the Rocky Mountains, is quite scenic and undeveloped. As a result, it is a magnet for groups and individuals seeking backcountry recreation experiences (i.e., hunting, off-road-vehicle use, hiking or backpacking) or simply seeking the solitude that the area offers. Consequently, the local economy has developed around a strong tourist trade.

Public response to the Forest Service proposal was quite predictable. Local residents, lodge and resort owners, and the Chamber of Commerce immediately expressed their opposition to the proposed road and logging. As reported in a regional paper:

. . . most of the residents in the summer-resort area, especially those owning nearby dude ranches, were convinced that large scale timbering would harm their tourist-based economy, which depends on clean water, abundant wildlife, spectacular scenery, and lots of solitude. (Wiggans, 1983, p. 6)

They initiated a petition drive, eventually soliciting 1000 signatures. Additionally, they rallied the support of state and federal legislators who in turn began barraging the agency with questions regarding their constituents' concerns. The support of state and federal environmental organizations was also obtained, a lawyer was hired, and lawsuit preparation begun.

The Forest Service reacted with surprise to the vocal and adamant opposition to its plans. They held public meetings with the different groups and made adjustments to their plans in ways that, they believed, responded to the concerns being raised. Their modifications only heightened the opposition, however, as the individuals and groups felt their concerns were not being addressed and perceived an agency with responsibility for public resources being unresponsive to the public interest.

As the dispute heightened and moved predictably towards administrative appeals and the courts, the suggestion was made that a mediator be secured to help resolve the dispute. The initial reaction on the part of all parties was that the dispute could not be mediated. The agency believed that the local residents merely wanted to bring its plans to a halt and the local residents felt that the agency would never budge from its proposal. Even the mediators, when first told of the dispute, were skeptical, thinking that it was a classic build–no-build conflict, leaving little room for agreement between the parties.

What happened, though, was a surprise to all parties, given the contention created through the Forest Service assessment and decision-making process. Two mediators from the Mediation Institute in Seattle, Washington agreed to facilitate discussions between the different groups. Twenty-six individuals, representing timber companies, ranchers, resort owners, recreationists, and environmental groups, participated in the mediation. These individuals represented those currently using the area, those who had been outspoken at earlier Forest Service public meetings, and/or those who the agency felt had a stake in whatever decision was reached. During a series of two, 2-day meetings, parties were able to reach agreement on 95% of their differences. They found, when finally sitting down together at the same table rather than shouting at each other in public meetings, that, in fact, they did have some common concerns. They found that they had misinterpreted some issues and concerns. They suddenly discovered that their assumptions about what the others wanted were off-base, that there were other ways of satisfying the different interests involved, and that the Forest Service was not indifferent to their concerns (Tableman, 1985).

Forest Service representatives attended the meetings in a support capacity. They provided maps, aerial photographs, and whatever other information was needed during the meetings to facilitate discussion and to develop and analyze alternatives. Additionally, they indicated when different proposed solutions would be legally, biologically, or administratively impossible for the agency to fulfill. Although the Forest Service did not (nor could they) agree to implement intact whatever agreement was reached in these meetings, they did make it known that they would seriously consider these recommendations in their administrative review process. As Forest Supervisor Paul Sweetland commented, ". . . if [what comes out of these meetings] is a consensus, these people have conveyed quite a message" (p. 27). However, although there were certainly incentives for the Forest Service and the other parties to comply with any agreement reached in order to avoid protracted delay, the agency was still bound to fulfill its procedural public review and comment requirements, using the agreement as their "proposed" decision.

The mediators facilitated the meetings in a manner encouraging discussion between the different groups. They used small-group discussions on the first day to help the groups clarify and articulate their concerns and needs about the area and then to elaborate upon the similarities and differences between their needs and concerns and those of the other participants. Eventually, these smaller groups began to explore ways in which the differences between them might be reconciled. By the end of the first day they were surprised to see how little distance separated them. On the second day, the larger group of 26 persons was consolidated into a smaller, more workable group of 11 persons through a self-selection process. A breakthrough in negotiations came when the area in question was subdivided into separate sections, representing different viewsheds and resource values. This smaller group was then able to come up with a specific proposal, eventually approved by the larger group.

By the end of the second day, the group had agreed to a compromise proposal for the Forest Service. Forest Supervisor Sweetland responded immediately that he could support 80% of their agreement but that he had problems with the remaining 20%. At that point, the second set of meetings was scheduled. Additionally, the Forest Service held a trail ride through the proposed road area for all participants. Discussions in these further meetings narrowed their differences to 5 percent. Forest Supervisor Sweetland issued his decision notice at this point, employing his professional judgment on the few remaining differences. Although not 100 percent of what the groups had originally advocated, they nonetheless felt that the decision was far better than what they could have achieved through appeals or the courts and therefore whole-heartedly supported it.

Through this mediation process, the participants found that there were ways for the agency to conduct its timber sales without harming the scenic viewshed, recreation amenities, or local tourist economy. The local residents and environmental groups were not opposed to timber harvesting anywhere, just in the section of forest providing a scenic backdrop and most accessible recreation opportunities to their community. Local ranchers did not promote the road proposals as most groups had presumed. Rather, minimal roads would provide access to their grazing lands without encouraging cattle rustling. Additionally, the local timber companies preferred logging in one section of the proposed harvesting area but not the area with the scenic viewshed. They preferred an area closer to their mills that would only require a 10-mile trip for their logging trucks, rather than the 100-mile trip if the more scenic area were logged.

The mediation process allowed the needs of each interest group, including the agency's, to be articulated, and alternatives to be developed to address those needs. These alternatives were then collaboratively evaluated and a joint-solution reached. Each interest group had a hand in developing the solution, and they could see how and where their concerns were being addressed, and why they might not be addressed in other areas. Each interest group had a role in making the critical tradeoffs. As a result, they supported the final agency decision and are now actively assisting the agency in implementing this decision.

THE WILLAMETTE NATIONAL FOREST CONSENSUS-BUILDING PROCESS

The Willamette National Forest is located in Western Oregon within a few hours drive of the state's three major cities: Portland, Salem, and Eugene. Because it is so close to Oregon's population centers, the Willamette National Forest is used heavily for recreation. Within the forest is a 20-mile-long, ¾-mile-wide area known as the South Fork Corridor, because it straddles the South Fork of the McKenzie River. It provides access for backpackers to an adjacent wilderness area and is, more generally, used for local hikes, picnics, and scenic drives. The area had been given the Forest Service's *Scenic I* and *Scenic II* classifications, meaning that the local Forest Service officials were required to accommodate some timber harvesting there while maintaining the area's special scenic quality.

In early 1982, the District Ranger developed a proposed 15-year timber sales plan for the South Fork Corridor. Two timber sales were planned for 1982. Jim Caswell, the District Ranger, knew the proposed timber sales would be controversial because of the area's heavy recreation use and its proximity to a wilderness area. As a result, he held a public meeting in March to discuss the proposed timber harvest with any interested groups and individuals. Subsequently, he organized a field trip for meeting participants to tour the proposed sites. As public interest and concern grew, Caswell held an additional public tour of the planned harvest sites in early June. As Ranger Caswell describes this tour:

> We spent the day looking at units, talking about the concept, showing them all the computer data, showing them that the proposed cuts couldn't be seen, and how we were going to put this mosaic together using the latest landscape architecture principles. The trip went real well. It was a nice day, and everybody, I think, had a good time. But the bottom line was that no one really wanted timber harvested there. They wanted us basically to leave that piece of country alone, regardless of what our plan said. (Mason and Desmond, 1983, p. 51)

Even with touring the site and having an indepth explanation and discussion of the Forest Service's plans, the participants did not accept the agency's rationale and continued voicing their opposition to the District Ranger's proposal.

When it became apparent to the Forest Service that opposition was mounting to the planned timber sales and that a number of different groups were mobilizing, the District Ranger tried a different approach. Following the public tour, a series of informal meetings and contacts occurred between Forest Service representatives and those groups and individuals opposing the agency's plans. Additional site visits were made, and, as specific concerns were presented, the District Ranger began making some adjustments to the Forest Service plans in an attempt to accommodate these concerns. Some planned clearcuts were changed to selective cuts and the boundary of another clearcut was shifted in order to protect a popular hiking trail.

These changes were not sufficient to satisfy the concerned hiking and environmental groups. Consequently, a group of about 20 individuals representing several different groups approached the Forest Service officials in hopes of pursuing a common ground. A citizens' task force on the issue was formed at the agency's suggestion, a smaller group created from the original 20 individuals. The citizens' task force and the Forest Service then began a 6-month effort during which time they, in their words, "moved from confrontation to compromise" (Mason and Desmond, 1983, p. 55). A series of proposals and counter-proposals were developed by the citizens and the agency and eventually a compromise proposal was reached. A breakthrough in the dispute occurred when the area was sectioned into five separate areas based on different resource characteristics. Each area was dealt with separately. One was reserved as an elk habitat with no timber harvesting planned. Two proposed timber sales were shelved until further planning could occur. A third timber sale was accepted by all the groups and has been scheduled.

This consensus-building effort provided the different groups, including the Forest Service, with an opportunity to focus on their direct issues of concern rather than those positions that might indirectly satisfy their concerns. It allowed the citizen participants an opportunity to creatively develop, rather than just provide input to, the different alternatives, and then to evaluate these alternatives with the Forest Service, rather than simply responding to plans set by the agency. As opposed to concentrating on what was wrong with the proposed decisions, the consensus-building effort provided a forum wherein mutually acceptable proposals could be devised and then supported. As one citizens' task force member commented:

> Many people went in with a rather negative feeling and came out with a very positive feeling toward both the product and the process. Most people say nobody is satisfied with a compromise, but in this case it seems that everyone is satisfied. (Mason and Desmond, 1983, p. 56)

In this case the impact assessment and decision-making process was amended in such a way that, should a common ground exist between the many interests involved, it could be found. Furthermore, the process changed the nature of the interactions between the different groups and the agency in a positive manner, continuing to facilitate collaborative discussions on other issues in the Willamette National Forest.

THE BRIDGER-TETON NATIONAL FOREST *AD HOC* NEGOTIATIONS

In early 1978, Getty Oil Company filed an application for a permit to drill with the USGS Office in Rock Springs, Wyoming for leases it held in the Fall Creek section of the Palisades Further Planning Area. Getty officials were not aware of the ongoing leasing battle in another section of the Palisades FPA (see Chapter 4). Getty's leases had been issued in

May, 1970, well before the area became part of the Palisades FPA during the Forest Service's RARE-II program.

When Getty's APD was forwarded to the USFS Bridger-Teton National Forest headquarters in Jackson, Wyoming, public notice was routinely made of the impending Forest Service review and recommendation. Sierra Club representatives immediately responded. They had just filed their first appeal of the Regional Forester's decision to recommend leasing in the Palisades area and they merely supplemented this appeal with a new paragraph noting their opposition to Getty's Fall Creek plans and their reasons the Forest Service should be prohibited from taking action on this APD until after a final wilderness decision was made.

Getty officials foresaw all too well the events that would likely follow. They were prepared to drill then and did not want to spend years in court before learning whether or not they would ever be able to do so. Uncertainty and delay would be very costly to the company, as evidenced by their experience in nearby Little Granite Creek where, 8 years and $1.5 million after filing their APD, Getty still did not know whether or not it would be able to drill there. In the Fall Creek case, rather than wait for the Sierra Club appeals to be concluded with no guarantee that the courts would not then become involved, Getty officials decided to play a more active role than traditionally adopted by APD applicants.

At Getty's encouragement, Phil Hocker of the Sierra Club, USFS Bridger-Teton National Forest minerals specialists and the Forest Supervisor and Getty Oil representatives met together at the Bridger-Teton National Forest headquarters in Jackson to discuss Getty's proposal, the Sierra Club's concerns and what could be done to bridge the two groups. Phil Hocker expressed the Sierra Club concern that the proposed drilling would be an intrusion into an area that was under consideration for inclusion in the National Wilderness Preservation System. The Sierra Club was concerned that such an intrusion would ultimately affect the Palisades FPA wilderness decision and that, on those grounds, it should be defered or denied.

Getty representatives countered with assurances that the access road to the Fall Creek wellsite could be maintained, reclaimed, and the entire site reseeded to mitigate any impacts. They assured Hocker that Getty would work under close Forest Service and Sierra Club supervision in developing and reclaiming the site. Discussion then proceeded about the specifics of road location, construction, and reclamation, and of how drilling activities would be conducted. Visits to the Fall Creek site were made. Agreement was reached setting forth the conditions under which Getty could drill its exploratory well at Fall Creek, how and where the access road and wellsite should be developed, and how road and wellsite reclamation should occur.

The dialogue established between Getty Oil, the Forest Service, and the Sierra Club over the Fall Creek well allowed all three groups to determine whether or not a common ground existed between them. Because Getty was willing to accommodate Sierra Club concerns and because Sierra Club representatives were willing to be upfront about their concerns with the proposal, the dialogue uncovered a mutually acceptable operating plan. Forest Service officials viewed the agreement as a godsend. They were relieved of the frustrating and time-consuming burden of themselves trying to resolve the differences between these two adversaries. Moreover, the negotiations between the Sierra Club and Getty are often extolled by Forest Service officials desiring similar, harmonious outcomes in other difficult cases, but unsure how to achieve them. Getty has since drilled its well and fully reclaimed its wellsite and access road to both the Sierra Club and Forest Service's satisfaction. In fact, both Phil Hocker and Bridger-Teton National Forest minerals specialist Al Reuter describe the reclamation as "excellent" and "a model for other reclamation activities" (Personal communication, November, 1981).

THE IMPORTANCE OF PROCESS IN RESOLVING DISPUTES

Although many of today's public land managers do not admit it, managing the national forests is inherently a negotiation process; it is no longer solely the professional silvicultural scientific management process that it more dominantly once was. Now the number of groups involved, their interests, and what they have at stake is far greater. Although, in most cases, the professional land managers try to internalize this inherent negotiation process—being arbitrators of sorts—by listening to the many demands and trying to balance them into reasonable, professionally acceptable compromise outcomes, their efforts seldom succeed. Unlike officially sanctioned arbitrators, acceptance of Forest Service decisions is not required of stakeholders. As indicated by Chapters 4 and 5, there are many avenues for potentially influencing forest management decisions apart from the agency's formal decision-making process.

The first question that decision makers should ask themselves when confronting a complex situation is not *What* is the proper allocation of resources in this situation?, or *What* should we decide?, but *How* should we make such a complex, difficult, and controversial decision? In other words, the first question should be a *process* question (*how* to go about making a decision) and not an *outcome* question (*what* should be decided). A process focus immediately raises questions such as what information do we need, who should be involved, where can we get the information needed, and, what are the likely problems we will encounter and how might we overcome them? A process focus encourages consensus-building and collaborative problem-solving among affected interests when the decisions to be made are complex and value laden, and when there are limits to technical expertise in reaching solutions. It focuses attention on the incentives under which different groups are operating and encourages the scale of analysis to be a manageable one. Moreover, with an appropriately structured process, viable decisions become possible. These decisions, instead of being contested and delayed through appeals and lawsuits are understood, accepted, and supported by the groups involved (Wondolleck, 1985).

As an illustration of these concepts in practice, consider the Forest Service's Roadless Area Review and Evaluation processes. If we knew in the early 1970s what we now know, the RARE-I and RARE-II processes would never have been structured as they were. The many groups affected by wilderness designation decisions had several *incentives*. First, they had great incentive to actively lobby the Forest Service during its review and evaluation efforts. A passive stance would certainly not encourage much agency attention to a group's concerns. Second, they had every incentive to then criticize the Forest Service's analyses and recommendations once completed. The old adage "the squeaky wheel gets the grease" holds true. For a group to accept agency recommendations when all other stakeholders are contesting them only meant that the group's interests might be sacrificed. Third, they had every incentive to (and, in fact, would have been foolish not to) then lobby Congress for allocations more favorable to their interests, regardless of what the Forest Service proposed. Congress, after all, had final decision-making authority over wilderness designations. These were the rational responses of each interest group because these responses were the only formal ways provided to influence wilderness decisions. It would have been more surprising if the groups had not become so embattled; complacency or accommodation was simply not in their best interests, given the incentives provided by the decision-making process.

Understanding this dynamic leads to several considerations in developing a suitable wilderness evaluation process. First, it should involve representatives of all affected groups in order to encourage compromise when it is possible. Second, it should involve those with authority to make the final decision; it should involve Congressional representatives, not just Forest Service officials. Third, the scale of the effort should be such that key interests can be meaningfully involved and the resource base under consideration be of a manageable size. Therefore, such a process would more likely deal with the wilderness designation question on a forest-by-forest or state-by-state basis.

Successful wilderness legislation in the late 1970s was scarce. Two key pieces that were passed, however, involved processes that include those considerations mentioned above. The late Senator Frank Church intervened in the Gospel-Hump Wilderness considerations in Idaho and encouraged the compromise settlement between the disputing parties that he then successfully introduced as legislation to Congress. Similarly, the late Senator Henry Jackson became involved in the Alpine Lakes Wilderness designation process in Washington State, leading to Wilderness legislation there as well. Furthermore, current wilderness designation efforts follow this same model. Wilderness evaluation and designation is occurring on a state-by-state basis, involving negotiations that include each state's Congressional delegation, the key interest groups, and the Forest Service. It is clearly not an easy process; nonetheless, it is a process that is finally leading to implementable decisions.

The approaches taken in the Monongahela and Jefferson National Forest planning processes described in Chapter 7 and the San Juan, Willamette, and Bridger-Teton National Forests described in this chapter applied these principles in practice in a formal yet *ad hoc* manner. Although involving (in the San Juan and Jefferson National Forest cases) professional mediators and, in all cases, processes incorporating the key elements of environmental conflict management, the processes were nonetheless separate from officially sanctioned agency administrative procedures.

The Forest Service has been encouraging, with some success, the use of environmental conflict management and dispute resolution concepts. Since 1982, they have contracted with ACCORD Associates of Boulder, Colorado (an organization specializing in environmental and natural resource dispute resolution) to conduct indepth training workshops for both line officers and staff in natural resource conflict management skills and techniques. Participants have been applying the skills acquired in these workshops in a wide range of situations, both internal to the agency as well as in disputes involving one or more national forest user groups or other government agencies. Nevertheless, at times these participants encounter difficulties because either their peers or superiors have not had the benefit of the training and hence do not understand its potential and are not supportive, or existing administrative procedure and direction does not accommodate such efforts when they would like to try them (Wondolleck, 1986). Although this training program represents the critical first step in reforming agency decision-making efforts, more formal institutional change is needed to support the efforts of those agency officials trained in conflict management skills and to potentially resolve those disputes that are now permitted to escalate.

Not all disputes encounter such favorable climates for resolution as those found in the three cases presented above or the few successful forest planning cases mentioned in Chapter 7. Individuals familiar with the potentials of formal conflict management efforts (as in the San Juan case) or having the wherewithall to confront the agency with such a proposal (as in the Willamette or Bridger-Teton cases) are not present in all national forests. Furthermore, agency officials are not always receptive to suggestions of more collaborative processes. Therefore, although incorporating the key process considerations discussed above may be sufficient for some agency situations, others may be better supported by a more formal structure that both sanctions and accommodates the effort.

The process described in Chapter 9 is built around the five objectives discussed in Chapter 7—building trust, promoting understanding, incorporating value differences, providing for joint fact-finding, and encouraging collaboration. This process provides a more formal structure within which some national forest management disputes might be resolved. The U.S. Environmental Protection Agency has successfully experimented with a process very similar to that to be described. (Harter, 1984b; Susskind and McMahon, 1985). The Forest Service might consider experimenting with it as well. Of course, less formal efforts that may anticipate and avoid full-scale disputes should not be downplayed should such a system be put in place. Indeed, if Forest Service officials apply conflict management principles in their everyday management of the national forests, many disputes may not escalate to the degree that triggers the more formal process proposed in Chapter 9.

Resolving National Forest Management Disputes

An Experimental Process

The process outlined in Chapter 9 is an experimental one. If pursued, it would provide the U.S. Forest Service and its constituent groups with the opportunity to try reaching agreement on selected issues that currently divide them and, in so doing, to determine the potential for resolving a wider spectrum of disputes in the agency. The procedural steps comprising this experimental process are designed to supplement existing Forest Service decision-making processes. They have been developed recognizing and reinforcing the statutory authority and responsibilities of the Forest Service in national forest management.

This amended process acknowledges both the judgmental and scientific aspects of many national forest management decisions. It acknowledges the critical need for Forest Service expertise and experience in making these decisions. It acknowledges as well, however, because of the value judgments inherent in these decisions, that professional expertise is not always enough in determining what decision should be made once the implications of various alternatives are understood.

The proposed process is designed to provide a forum within which groups can more directly ensure that their concerns are addressed in decision-making. Although it in no way precludes administrative appeal or judicial review of decisions made, this process should encourage participation, because it provides a more direct and manageable outcome as opposed to the uncertainties and costly delays associated with the current process. In this sense, the process has built-in incentives for participation; each group's power should be greatest *in* the process

225

rather than, as is currently the case, by following other avenues for intervention.

THE U.S. ENVIRONMENTAL PROTECTION AGENCY EXPERIMENT

The process outlined below is designed after other similar experimental efforts now underway in several federal agencies, including the U.S. Environmental Protection Agency (EPA) (Harter, 1984b; Susskind and McMahon, 1985). Since 1983, the EPA has used what they refer to as a *negotiated rule-making process* to bring together those groups with a stake in the agency's regulatory decisions. The EPA discovered that, of the 300 regulations that the agency was issuing each year, 80% were ending up in court and 30% of those were being significantly changed as a result of the legal challenge. Agency officials estimated that 125 staff-years were assigned annually just to manage these lawsuits. An additional 25 person-years were consumed by Department of Justice staff and 175 person-years by plaintiff's counsel (Susskind and McMahon, 1985). The agency realized further that when these lawsuits failed their advocates would then simply return to Congress hoping for more favorable laws that would then commence the litigous cycle once again (Stanfield, 1986, p. 2764). In short, their predicament exhibited some striking similarities to the predicament faced today by the Forest Service (as well as by many other federal agencies).

Although acknowledging that the nature of their task would always promote conflict, EPA officials wanted to minimize the costly disputes whenever possible. Therefore, they began a formal, though experimental, negotiation process to see when and how accomplishing this objective might be possible. As current EPA Administrator Lee Thomas remarked in May, 1986, at the Conservation Foundation's Third Annual Conference on Environmental Dispute Resolution:

> Now I know we will never see the day when we will be involved in no litigation. I'm not sure I really want to see that day. After all, we deal with some very tough, very complex, and very controversial issues. Often, we are breaking new ground. When you are on the cutting edge, as we are, there is some comfort in knowing that your decisions are subject to judicial scrutiny. I consider it a form of quality control in public policymaking.
>
> The real problem is not with litigation *per se*, but rather with litigation as a predictable part of the standard rulemaking process. Once that happens, and it is happening all too often, you find the federal system of checks and balances being disrupted. In effect, you transfer responsibility for federal rulemaking to the judicial branch. I'm not comfortable with that system. I don't think Congress ever intended this to be the case.
>
> We have in place comprehensive air, water, toxics, and waste manage-

ment regulatory programs and the enforcement tools needed to make them credible. More than anything else, we now need to find ways of making these programs work better, faster, and more completely. Lawsuits don't usually do that.

At the heart of regulatory negotiation, ("reg-neg") is one simple goal: a better regulation, arrived at by listening to and talking with those most directly affected by it, and no litigation. Instead of spending time and money fighting our way through the courts, "reg-neg" can allow us to invest those resources in implementation and enforcement. And it is from carrying out the regulation itself that we will see environmental progress.

By mid-1987, five rules had been successfully negotiated and implemented without appeal or lawsuit and two additional rule-making negotiations were underway. Following the same series of procedural steps that I am proposing, agency officials found that this experimental route markedly shortened the time involved in EPA promulgation of a rule. It traditionally takes a minimum of 3 years to finalize a rule through the agency's normal rule-making process. In this traditional process, EPA staff develop a proposed rule and then publish it in the *Federal Register* for comment. After receiving "hundreds or thousands" of comments on this proposed rule from interested individuals and organizations, the staff evaluates these comments, finalizes the rule and awaits the almost inevitable lawsuit. In contrast, the rules negotiated in the EPA's regulatory negotiation process have elicited less than 20 comments each and have not been legally contested. According to Chris Kirtz, director of the EPA's regulatory negotiation project, "The amount of time it takes to go from proposed to final rule and the amount of staff work [during that stage] is infinitesimal" (Stanfield, 1986, p. 2765). Moreover, according to Robert Ajax, chief of the EPA's air program standards development branch, "The process produced a better rule than the agency would have been able to get out [in the traditional process]" (p. 2765).

Implementation of final rules is facilitated as well because of the consensus between those who must comply with it (i.e., industry) and those enforcing it (usually the states). Implementation is further promoted because lawsuits have been avoided. Furthermore, the new rules go into effect up to 2 years before they otherwise would have in the more adversarial system, presumably promoting environmental quality objectives in a more timely fashion.

Obviously, these successful rule-making efforts succeeded in large part because there was room for agreement among the parties and the process provided the forum and the structure to find that agreement. As one observer of the EPA process noted:

> The trickiest part, probably, is picking a suitable issue. It must be important enough to warrant the time and energy involved, but not so major that it strikes at fundamental principles that are non-negotiable. The controversy

> must be focused on definable points that are underpinned by facts all sides can agree on. There must be a clash of several viewpoints—if not, why bother negotiating—but to make the whole thing manageable, these divergent views must be represented by a few well-identified, preferably organized, interests. (Stanfield, 1986, p. 2767)

Furthermore, all parties had incentives to reach agreement quickly rather than continue to delay. The risks and costs of delay were deemed undesirable by all parties. As well, each rule-making effort was systematically structured and facilitated, at times involving outside professional facilitators and at other times agency officials with facilitation skills. Two, full-day sessions were held once a month over a 6-month period in the "typical" negotiated rule-making process. Between these formal meetings, less formal small group caucuses and working sessions occurred with EPA staff assistance, as well as numerous phone calls allowing parties to confer and bargain. EPA staff were busy at all times completing the technical and analytical tasks critical to any rule-making process.

On the downside, the EPA has found that these negotiations are more staff-intensive than expected. However, according to Fiorino and Kirtz (1985):

> nearly all of the participants in the negotiations concluded that the process worked better and yielded more acceptable regulations than they might have expected under conventional rulemaking. In addition, participants pointed to improved understanding of technical issues, fuller appreciation of the institutional positions of the other parties, and an awareness of the potential for negotiation as an alternative to the standard method of rulemaking. (pp. 29–30)

Susskind and McMahon (1985) similarly found that participants believed the negotiated outcome "was far better than what they might have expected had they gone to court" (p. 151) and "reduced the amount of time involved in litigation and subsequent administrative wrangling" (p. 151).

APPLYING THE REGULATORY NEGOTIATION EXPERIENCE TO NATIONAL FOREST MANAGEMENT: A PROPOSED U.S. FOREST SERVICE EXPERIMENT

Several criteria were mentioned in Chapter 8 that make national forest management disputes amenable to resolution. The different parties having a stake in the decisions to be made are well known and organized. The issues of dispute are well-defined. Power between these parties has become well-developed and somewhat balanced through lawsuits, administrative rulings, and Congressional mandates. These management and planning decisions inevitably must be made. The different parties to the disputes have become frustrated pursuing other avenues yet fail to obtain representation to their satisfaction. It is costly to all parties, the Forest Service included, to continue in an adversarial process, never focusing on or resolving the real issues of concern (Harter, 1982; Raiffa, 1982). However, merely satisfying these theoretical criteria does not then pave the way to successfully resolving a particular dispute. If these elements are present in a conflict situation, the next step is structuring a process to potentially resolve the dispute, a process that involves all key stakeholders, has the trust of its participants, and encourages collaboration and negotiation (Susskind and Weinstein, 1981). The following process has been developed with understanding of both the nature of national forest management conflicts and the lessons of past efforts, such as those at the EPA, to manage conflicts while developing viable decisions and programs.

Susskind and McMahon (1985) have identified the three key stages comprising the EPA's regulatory negotiation process, each of which contain critical steps towards ensuring that the process will be trustworthy, representative, and effective at reaching acceptable outcomes when possible. These three stages are: prenegotiation, negotiation, and postnegotiation. The rest of this chapter sketches the activities of each of these three stages as applied to national forest management disputes. The intent of the chapter is to give a sense of what such a process would entail in a Forest Service application, key issues to consider, and how to begin. It is not meant to be an operating guide through the process, however. What follows is simply an introduction to the idea and its potential application in a Forest Service setting. Should Forest Service officials choose to experiment as the EPA has, the assistance of convenors and those familiar with the inner-workings of the regulatory negotiation process in other federal agencies should be arranged.

THE PRENEGOTIATION STAGE

During this first stage of the amended process, the Forest Service must decide whether or not to initiate the experimental process and, if so, to employ a convenor to help identify representative participants. The convenor does the background work for the process, including meeting with stakeholders, encouraging participation, drafting preliminary groundrules, and providing assistance to potential participants in obtaining financial or technical resources or negotiation process skills training (Susskind and McMahon, 1985).

TRIGGERING THE PROCESS

Not all national forest management decisions are controversial; not all processes yield decisions that are not viable. Because this process is merely supplementing existing procedures, it needs to be *triggered* when those decisions arise that will potentially benefit by it; decisions that exhibit those characteristics highlighted in the cases described in Chapter 4. Because this process would initially be an experimental effort, not formally institutionalized in agency procedure, the Forest Service might select several potential issues with which to experiment through dialogues with various national forest user groups. The EPA solicited recommended rules from the industries it regulates and from concerned interest groups, and then from this list chose those with the greatest promise.

Harter (1984a) has identified eight conditions that make a situation "hospitable" for these direct negotiations:

1. There are a limited number of interests that will be significantly affected, and they are such that individuals can be selected to represent them;
2. The issues are ripe and mature for decision;
3. The resolution of the issues presented will not require any interest to compromise a fundamental tenet or value, since agreement on that is unlikely;
4. There is a reasonable deadline for the action so that unless the parties reach an agreement, someone else will impose the decision;
5. There are sufficiently many and diverse issues that the parties can rank them according to their own needs and priorities;
6. There is sufficient countervailing power so that no party is in a position to dictate the result;
7. Participants view it as in their interest to use the process as opposed to the traditional one; and
8. The agency is willing to use the process and will appoint a senior staff member to represent it. (p. 1393)

Harter argues that such negotiations should involve no more than 15 participants, but Susskind and McMahon (1985) have found that 25 participants has been workable in the EPA experience, and that some processes may be able to accommodate even more. The process is designed to provide a means for facilitating, not delaying, decisions. It does not represent more work for the agency or other participants; it simply redirects efforts that already occur. Hence, the incentives built into the process should be such that only those cases that can potentially be resolved will trigger it.

Forest Service decision-making is generally of two types, occurring at two levels in the agency's hierarchy. The first type of decision involves policy affecting agencywide administration. The Washington, D.C. Office establishes policy. It structures how the agency will implement Congressional mandates by developing guidelines, procedures, and regulations that are then placed in the *Forest Service Manual*. An example would be the Further Planning Area (FPA) Stipulation and Guidelines that played a critical role in the Palisades leasing dispute discussed in Chapter 4.

The second type of decision occurs when officials in the regions, forests, and districts apply these internal mandates to site-specific proposals or general management tasks. Implementing the FPA stipulations and guidelines in the Palisades case, or permitting procedures in the Cache Creek and Little Granite Creek cases are examples of site-specific decision-making. Because disputes arise in both policymaking and policy implementation, attempts to resolve them should also occur at both levels.

The Assistant Secretary for Natural Resources and Environment in the Department of Agriculture might trigger this process for policy-level disputes. (The Assistant Secretary has jurisdiction over the Forest Service.) The Forest Service Chief might trigger it for site-specific disputes. These two individuals are chosen because they do not have initial responsibility for making the particular decision, although that responsibility could be appealed to them by dissatisfied groups. Therefore, they will not be party to the dialogue that ensues. They are in a position of authority over those in charge of making the final decision. As such, their action legitimizes the process. Moreover, by institutionalizing this process, the Assistant Secretary or USFS Chief empower the different groups participating in it and officially sanction the effort.

The Assistant Secretary or USFS Chief triggers this process upon petition of the Forest Service Chief (for policy-level decisions) or the responsible Regional Forester (for site-specific cases). A particular public land user potentially affected by a decision can also petition to have this process triggered.

The process is triggered when the Forest Service Chief or Assistant Secretary files a notice of intent in the *Federal Register* to commence this process and convene the different participants. Consistent with current review procedures, this notice should be supplemented by announcements sent to those on the agency's already-established mailing list of interested parties and to the media.

To a large extent, the cases that might benefit from this process are self-defining. When the Washington Office develops rules governing decision-making, officials there generally have a good idea at the outset what different groups are going to think about these rules and the concerns they will express. At this point there are few surprises. Similarly, when, for example, a lease or permit application is filed or a timber sale is proposed, the District Ranger or Forest Supervisor has a good idea who is going to be concerned and what the responses are likely to be.

Decisions that should trigger this process are those in which different interests clash and in which much is at stake; decisions that potentially affect some user groups beneficially and others adversely and therefore are likely to be opposed, regardless of outcome. For example, should exploratory oil and gas drilling or timber harvesting be proposed in an area with valued scenic resources, important wildlife habitat or wilderness characteristics, this process might be commenced. In so doing, the critical issues of concern can be identified early in the process. Moreover, if there is a mutually acceptable outcome, it can be determined before the predictable battle lines are drawn.

Policy decisions broadly affecting the management of specific areas—wildernesses, further planning areas, wild and scenic rivers—should trigger this process. More generally, decisions that will likely be contested at later implementation stages if disputes are not resolved might follow this process. Similarly, cases warrant triggering where initial announcement of intent to develop a rule or regulation or evaluate a lease or permit application generates conflicting input. And, precedent-setting decisions broadly affecting a range of different users in numerous decisions over time should be subject to consensus-building in order to avert the domino effect, wherein they are contested when applied in case-after-case.

Acknowledging that not all rules are amenable to negotiated agreements and, moreover, that the structure of the negotiation process and who is involved in it are key to its potential success, EPA staff have outlined the following recommendations for other federal agencies undertaking similar processes. These recommendations highlight those types of issues that are most likely to succeed, as well as who should participate in the process:

- The proposal should require the resolution of a limited number of interdependent or related issues. There should be several ways in which the issues can be resolved. The relevant legislation should accommodate these alternative outcomes. There should be no serious obstacles to implementing a negotiated solution.
- To promote timely resolution and to limit a participant's ability to

gain from delay, there should be a legislative or judicially imposed deadline or some other mechanism forcing publication of a rule in the near term, (i.e., 8 to 12 months).

- Some or all of the parties should have common positions on one or more of the issues to be resolved that might serve as a basis for additional agreements during the course of negotiations.
- The costs and benefits should be narrowly concentrated on a few identifiable entities.
- Those participants interested in or affected by the outcome of the development process should be readily identifiable and reasonably few in number. They should have sufficient resources to take an active role in negotiations, and should have relatively equal power to affect the outcome.
- The parties should be likely to participate in negotiations as an alternative to litigation. They should feel more likely to achieve their overall goals using negotiation rather than existing alternatives.

CONVENING REPRESENTATIVE GROUPS

Upon triggering this process, representatives must then be selected to participate in it. The Assistant Secretary or USFS Chief's *Federal Register* announcement indicated their intent to commence this process by convening the different groups. The announcement invited concerned groups to participate in *selecting* the participants. As seen in the San Juan and Willamette National Forest cases discussed earlier, if every individual were to participate the process would be unwieldly and unlikely to succeed. Discussions were facilitated when representatives were self-selected from among the larger groups to participate in more intensive discussions and negotiations.

In the EPA negotiated rule-making processes, stakeholders were identified and representatives self-selected through the efforts of an independent convenor. A convenor could serve the same function for the Forest Service, at the request of the Assistant Secretary or Forest Service Chief. The convenors used in the EPA processes were professionals specializing in environmental conflict management and having some professional facilitation and mediation skills. There are many organizations nationwide offering assistance in resolving environmental disputes, any of which might be a candidate to convene a group. Other potential convenors might be members of the Federal Mediation and Conciliation Service, should it be willing to take on disputes of this nature, or the American Arbitration Association.

It is important to understand that the convenor does not select the participants. Instead, they assist the groups in selecting their own representatives: people who are trusted and respected, whose decisions will be supported and accepted, and who will be able to speak for the groups they are representing. Throughout the process, these participants maintain contact with their constituencies. By this time, the national and regional environmental and industry groups are well-organized and identified, thereby facilitating convening their representatives. The Assistant Secretary and Forest Service Chief may assist the convener in bringing together the representatives of other government agencies when their participation is warranted. The convener not only brings representatives of various stakeholding groups together but also assists them in understanding the process, identifying concerns and coordinating efforts between participants within a group.

The convenor's task of coordinating efforts within the different groups is critical in order to avoid disjointed or inconsistent negotiations. For example, when the Sierra Club sued the Forest Service over leasing in the Palisades Area of Wyoming and Idaho, U.S. District Court (Washington, D.C.) Judge Aubrey Robinson suggested that the groups try to settle their differences out of court. He gave them 1 month to reach a settlement before he would commence judicial proceedings. Although negotiations initially were encouraging, disagreements and uncertainty on the part of the government representatives (attorneys for the BLM, DOI and USFS) helped to preclude any final agreement. The government representatives did not have power to make the decisions necessary to an agreement and, therefore, had to continually go back to their respective agencies before making any commitments. They also did not have expertise in the areas under consideration and needed to consult with other agency officials for information. Additionally, they had not developed any consensus on a consistent negotiating strategy at the outset, so they actually represented two bargaining teams, not one. Although the lease applicants and the Sierra Club seemed able to reach agreement on an outcome that would accommodate their concerns to their satisfaction, the problems among the government representatives precluded this outcome.

Three factors are critical in convening representatives of interested or affected groups. First, participants must represent all interests at stake. If they do not, then an overlooked group will not necessarily support any agreement reached. For example, when leases were originally filed in the Palisades area in 1977, the Forest Service was in the midst of its wilderness reviews. It initially decided to grant the leases but the Sierra Club appealed the decision. Upon appeal the Regional Foresters in Regions I and II reached agreement with Sierra Club representatives not to issue any leases in wilderness study areas until the wilderness review was completed (Nelson, 1982, p. 45). Because oil and gas industry representatives were not party to these negotiations, however, they did not feel bound to the agreement reached. As a result (as discussed in Chapter 4), the Mountain States Legal Foundation took the Forest Service to court to force them to make the leasing decisions (499 F.Supp. 383(1980)).

The second critical factor in convening participants is to ensure that the representatives have decision-making authority within their respective organizations. This requirement is critical to avoid the problem mentioned in the Palisades case where government representatives had to keep going back to the agencies for information and direction. Third, participants must have the trust and respect of the constituent groups they are representing in order to make decisions that the group will support. If this trust is lacking, there is nothing preventing a splinter group from forming to question any agreement reached.

THE NEGOTIATION STAGE

Once the participants are convened, this process essentially becomes an amended version of the current environmental impact assessment process. As designed, this process satisfies NEPA requirements for assessing the impacts of several different alternatives before making a decision. The Forest Service may want to consult with the Council on Environmental Quality to determine whether or not scoping requirements will be satisfied in this process, or if an additional public meeting should be held to accomplish the required scoping objectives. This meeting might help the participants ensure that their initial agenda has not overlooked any critical issues. However, should these participants have been successfully selected at the outset to represent all affected groups, scoping is built into the fact-finding and issue-identification stages and an additional scoping meeting would likely raise few, if any, new issues. Susskind and McMahon (1985) list the following tasks for the participants during this negotiation stage:

> The parties need to structure a work program, agenda, and timetable; undertake a process of joint fact finding, using consultants if necessary; organize subcommittees or working groups to develop preliminary drafts of proposals; confront major differences in their interests; produce careful summaries of the agreements reached at their meetings; and prepare a written draft of the proposed rule for all the participants, including the agency, to circulate for review. (p. 150)

They found one of the key conditions satisfied by the EPA, fostering the success of its experiments, was that the agency gave all participants an opportunity to shape the scope of the negotiation agenda and to jointly select a facilitator. Additionally, they gave all participants access to all the information they requested.

FACT-FINDING

The intent of the Negotiation Stage is to more directly incorporate the concerns of all potentially affected public-land users in the environmental assessment process. The first stage the Forest Service conducts in the current process is acquiring information with which to make their decision. In this amended process, facts are similarly compiled, but in a more participatory manner. Meetings are held during which agreement is reached between all participants on the scope and boundaries of the assessment. Once these factors are defined, attention turns to acquiring the information needed before a decision can be made. For example: What resources exist in the area? What are likely impacts on these resources and how might they be mitigated or avoided? What different means are available to satisfy the proposed objectives? Who will likely be affected both beneficially and adversely by the proposal and how, specifically, will they be affected? The EPA established a $50,000 "resource pool" at the outset of each of their negotiated rule-making processes to be used by all participants in obtaining information and covering other costs associated with their involvement in the process. In addition to supporting the group's data collection and analysis efforts, establishment of this resource pool also illustrates the agency's commitment to the process and allows participants to acquire nonagency studies when warranted.

At this point, the participants might want to jointly select an independent facilitator or mediator to assist them through the negotiations. The original convenor could assist the groups in jointly selecting a facilitator or mediator if desired, or could become the formal mediator him- or herself, as was the case in some EPA processes. In other cases, EPA officials whose positions were not associated with the rules being negotiated, successfully served as the facilitator. There are numerous organizations and individuals nationwide offering mediation and facilitation services for environmental disputes.

Should the groups believe that independent analysis is needed, private consultants can be hired to address a specific question. The process is iterative: should fact-finding reveal unforeseen concerns, additional data may be desired and added to the agenda (Susskind and Weinstein, 1981). But, as in the San Juan and Willamette National Forest cases, these determinations are made together by the participants, not by an individual group.

IDENTIFYING ISSUES OF CONCERN

Each participant needs to identify the key issues of concern for his or her organization given their understanding of the proposal and its implications. The point of the proposed process is to identify and then focus attention on the specific issues of concern before they are transformed into positions that then overshadow the actual concerns. (Fisher and Ury, 1981; Raiffa, 1982) If there is room for a "win–win" outcome (an outcome that is creatively molded to address and to accommodate the concerns of all parties), this interactive process is designed to determine what that outcome might be.

By breaking these key issues down into their component parts, the participants should be better able to identify and create alternatives that specifically address these issues. For example, in the San Juan mediation described earlier, rather than arguing for or against the Forest Service's road, each group identified its needs from the area. This approach led to recognition of the scenic backdrop of the area, cattle rustling, distance to local mills, and public involvement in road development as the key concerns to be considered in developing a solution.

Each representative has an incentive to be upfront and thorough about his or her group's concerns. The point of this process is to provide all affected parties with an opportunity to directly contribute to a final decision that accommodates their concerns; it does not preclude the other options. By not participating forthrightly, honestly, and reasonably in working toward a mutually agreed upon solution, a group merely throws itself back on the mercy of an uncertain appeals, judicial and legislative system. Moreover, it is not clear that judicial and legislative avenues will be as responsive if a means for settling a difficult dispute were not given a fair chance. Furthermore, because it is likely that groups will participate together more than once in this process, failure to negotiate in good faith only makes future negotiations more difficult, with other parties less likely to be reasonable and compromising. Because the process is a voluntary one, if a good faith effort to find a common ground proves unsuccessful, the parties can, at any time, return to the adversarial arena.

DEVELOPING ALTERNATIVES

One key task during this negotiation stage is to develop a list of mutually acceptable alternatives to be analyzed and discussed. Developing these alternatives can be best described as side-by-side problem solving. Rather than selecting a standard range of alternatives, the alternatives generated in this process specifically address the issues raised and are developed cooperatively and creatively, as happened in the Bridger-Teton, Willamette and San Juan National Forest cases. As discussed in Chapter 3, the Forest Service frequently requires different mitigation measures or applies different stipulations to specific management activities. These measures either completely avoid or minimize potential impacts to surface resources. In the proposed process, these mitigation measures are evaluated and applied in a manner that is visible for all to see, and in a manner that ensures each group of its intent. In so doing, it should avoid any distrust or misunderstanding of the conclusions reached.

Because of the uncertainty involved in some forest management situations, contingency agreements might be considered when alternatives are developed. These agreements (if this happens, then that must be done) help offset the uncertainty that now makes decision-making so difficult. Additionally, it ensures an ongoing review process. Several court decisions and Interior Board of Land Appeals rulings indicate that it is a valid approach to decision-making. Some agreements might provide for the ongoing involvement of different groups in a monitoring or advisory capacity, as seen in the San Juan National Forest case.

In creatively and cooperatively developing these alternatives, other options that are usually not considered can be raised. If, for example, a proposed oil and gas exploration raises concerns about the impacts to a town's streets or public services, then compensation to offset these impacts might be considered. Similarly, if a national forest recreation area or campsite is affected, compensation for relocating or mitigating the impacts might be considered. Different forms of compensation have been considered in resolving many environmental disputes (O'Hare, 1977; O'Hare, Bacow, and Sanderson, 1983). For example, in agreeing to an alternative that would both allow construction of the Grayrocks Dam in Wyoming and protect the endangered Whooping Crane's habitat downstream, the energy consortium involved agreed to establish a $7.5 million trust fund for maintaining and protecting the habitat (Wondolleck, 1979). Although such solutions can, at times, more directly address the issues of concern, the current process and judicial reviews seldom raise or consider them. Moreover, these traditional forums frequently preclude finding such solutions.

AGREEMENT ON A PROPOSED DECISION OR RULE

When specific issues of concern have been identified, tradeoffs must be made in order to achieve consensus. This final phase of the negotiation stage is perhaps the most difficult of the process. As discussed on page 236, a mediator or facilitator might be involved, at the request of the participants, to help them focus their discussions and develop possible agreements. The use of sample agreements drafted by the different participants or single negotiating texts help to focus discussion at this point.

To successfully reach an agreement may require expanding the list of negotiable items (Raiffa, 1982). Because many disputes are interrelated, tradeoffs can encompass factors in other disputes. Many participants undoubtedly will be negotiating together on several occasions. If participating in a similar process involving a different issue or land area, a final agreement may involve aspects of projects other than that being considered. Oil and gas lease applications are often combined on an areawide basis (perhaps through the forest management planning process). By expanding the context for decision-making, potential tradeoffs are increased and negotiations facilitated. Moreover, by consolidating leasing decisions in this manner, disputes involving a specific area can be resolved comprehensively rather than incrementally, on a lease-by-lease basis. Any agreement reached would cover where leases may or may not be issued and to what stipulations they should be subject.

This process will not always result in an agreement. In some cases participants may, after reaching partial agreement, leave the final decision to a jointly-selected arbitrator or to the Forest Service. Forest Supervisor Paul Sweetland took the 95% agreement reached in the San Juan mediation and made his decision. The affected groups accepted his judgment on the remaining 5% differences. The late Senator Frank Church (D-ID) similarly made a decision on the remaining points of contention in negotiations over Idaho's Gospel-Hump Wilderness Area to the satisfaction of all participants.

If agreement cannot be reached, it may indicate that the particular dispute is not suited to administrative resolution. The participants will then continue pursuing their concerns through traditional avenues. In such cases, if good faith efforts have been made to resolve the dispute, Congress or the courts may be more willing to adopt the issue and try resolving it in their forums. Moreover, the results of the assessment thus far should provide a more coherent and concise set of facts and issues with which a more substantive, rather than purely procedural, decision might be made. Furthermore, all participants should have acquired a broader understanding of the particular issue and its ramifications, potentially helping them to resolve other similar issues. Additionally, the process should help smooth the way for future collaborative interactions between the same parties (Buckle and Thomas-Buckle, 1986).

THE POSTNEGOTIATION STAGE

Because the authority for making national forest management decisions rests with the Forest Service, Forest Service officials cannot merely adopt the agreement reached in this process. They must first provide an opportunity for those not directly participating in these negotiations to comment on the proposed decision or rule. Therefore, the Forest Service must complete its procedural obligations and ensure that all concerns have been raised and addressed. Additionally, participants in the process must ensure at this point that their respective constituencies understand the proposed decision and support it.

As is customary, the Forest Service should place this *proposed* rule or decision recommendation in the *Federal Register* for comment, or should issue a proposed decision notice. When required, a public hearing should be held to obtain further feedback. The agency then accommodates these additional concerns in its final decision or recommendation. If considerable additional comments are received that are contrary to the proposed decision or rule, agency officials may want to reconvene the original participants in order to discuss the accommodation of these new concerns. The decision may be contested if all participants do not understand and agree to the changes that were made. A fine-tuning review may avoid this consequence. However, if the process were successfully designed at the outset to be representative and have constituent support, this outcome should be the exception, not the rule; the proposed decision should represent the collective interests of the participant groups and, thereby, public land interests in general. Susskind and McMahon (1985) found that one key to the EPA's success has been the agency's willingness to give participants ample opportunity to review the draft agreement and any subsequent comments and criticisms arising in the formal public review process, as well as to give each participant the opportunity to "sign off" on the final version of the agreement.

This final adjustment to the proposed decision should not reinstitute the problems that currently plague decision-making. Although the Forest Service was a party to the negotiations that originally developed this proposal, they have no incentive to drastically revise the proposal regardless of the fact that they have the authority to do so. Such an action voids the integrity of the process, ensures opposition to the decision, and taints the inevitable future interactions of the same groups.

FOREST SERVICE IMPLEMENTATION

Within this process, the Forest Service implements its decisions as outlined in current procedures. The agency's authority is retained. It fulfills its Congressional mandates, perhaps to an even greater degree if the outcome is "harmonious" as required in several land management statutes. The professional expertise that is critical to proper national forest management is not sacrificed to a purely political solution. Instead, the scientific expertise is applied to facilitate and inform the inevitable value judgments that must be made. (Of course, many Forest Service leasing and permitting decisions at the site-specific level actually constitute recommendations to the BLM or USGS. In these cases, the BLM or USGS, having participated in the process as government agency representatives, would have little reason to undermine the agreement and not abide by the Forest Service recommendation. Forest Service recommendations to the USGS or BLM are generally accepted as issued in the current process, even when the BLM or USGS has no role in their development.)

PROSPECTS AND POTENTIAL PROBLEMS

There are many reasons why an agency like the U.S. Forest Service might adopt conflict managing steps in its decision-making processes. The foremost reason would be to rebuild trust in the agency and its decisions among the many forest user-groups, and, in so doing, to establish a positive rapport and ongoing dialogue with these groups. Additionally, integrating affected groups into agency decision-making in this way will help build cooperation rather than adversarial battling among these groups, and will avoid the image tarnishing battles that now afflict the agency. It will help build a better public understanding of the forest management process, various groups' needs, and the constraints, as well as possibilities, of decision-making. It will encourage creative solutions to land management problems and is more likely to achieve decisions that are supported and respected than contested and ridiculed as in current processes. Furthermore, processes premised in conflict management principles will be more responsive to public desires and needs, while at the same time lending support to agency officials in bringing values and judgments to bear when professional expertise falls short. Successful processes, rather than diminishing agency authority, should increase agency independence by avoiding the political intervention that occurs when groups impose external pressures because they perceive the agency's process to be closed or unresponsive. A conflict management process such as that proposed here acknowledges that the agency is confronting a tremendously complex management task, and that there is no single "right" answer to what decision they should make. It is designed to facilitate decision-making by providing specific phases of analysis that directly build upon each other. In so doing, it should generate confidence in the conclusions reached and agreement on the value judgments and tradeoffs that must be made. This process is responsive as well to the current Congressional and administrative calls for regulatory reform.

Some political scientists might argue that the problem exists today because too much opportunity for public participation has been provided. They would urge a return to "responsible governance" and the cultivation, rather than the ridicule and demise of the public service professions (Alan A. Altshuler, personal communication, February, 1983). The process proposed here, however, is not meant to question the public forestry profession and to replace it with a purely political decision-making body. Instead, it is designed to assign the profession those tasks for which it has expertise and should prevail. And, in those cases where Congress has mandated tasks that require judgments beyond this scientific expertise, the agency will receive assistance by those directly affected by the decision.

The Forest Service's authority and responsibility for decision-making remains intact in this process. Moreover, if the process proves successful, the agency's position is strengthened. After two decades of controversy over its decisions, the Forest Service's image is badly tarnished. It now has few constituent supporters. It has often been said that a strong conservation-oriented Secretary in the Department of the Interior is all that is needed to effect a transfer of the agency from the U.S. Department of Agriculture to the Department of the Interior where the other land management agencies reside (Steen, 1976). It is a move that some interests advocate (Steen, 1976, Chapter 5, note 57). Should the Forest Service be able to rebuild its image and regain constituent support, however, then its position as a relatively independent administrative agency would be strengthened, not weakened by the process. The proposed process acknowledges the critical need for the expertise and experience of Forest Service officials and staff in making these controversial decisions. It should lend more credibility to and support for their analyses. In so doing, the process is structured to ensure that final decisions are based upon scientific land management needs rather than questions of administrative procedure or law as in the current process. As a result, the Forest Service can put greater effort into other land management tasks that are now shelved, while controversial decisions drag on.

This process is not solely advantageous for federal agencies responsible for public land management decisions. There are also incentives for environmental and industry groups to both advocate and participate in a consensual decision-making process. The uncertainty and delay now plaguing decision-making makes the current process very costly for industry. In 1983, 4 years and $1.5 million after filing its APD for Little Granite Creek, Getty Oil Company did not know whether or not it would ever be able to explore its leasehold. Getty officials would certainly have preferred knowing before making this expense whether or not it would ever receive a drilling permit. Environmental groups such as the Sierra Club and The Wilderness Society similarly find the established process costly.

For this process to be viable, two important factors must be satisfied. First, representatives from both noncommodity interests such as environmental groups and commodity-resource interests must participate with the Forest Service in fine-tuning the requirements and objectives of each step. All groups must be involved at the outset in order to understand this process, and to establish the ground rules that will govern how the process proceeds. If all groups are not involved initially and at this level of discussion, the possibility exists that one or another interest group will not trust the agency's motives in proposing it, and will therefore not participate. Second, a conflict-management capability must be institutionalized within the agency. When a dispute arises that triggers this process, a Forest Service administrative office housing a trained mediator or facilitator should exist in order to usher the dispute through the process. Leaving this responsibility to existing Forest Service staff not trained in conflict management would neither ensure that appropriate action is taken, nor provide the independence necessary to create a credible and trusted process.

Experimenting with the process proposed here is not without some administrative problems, in addition to those already mentioned. How, for example, will this process be financed? How will the Federal Advisory Committee Act provisions be fulfilled in convening these groups and encouraging an effective dialogue between them? Another problem is overcoming the immediate and inevitable distrust of the use of a bargaining process to resolve environmental disputes (Wondolleck, 1979; Susskind and McMahon, 1985). The mere suggestion that these public land management disputes might be managed and resolved inevitably provokes several, predictable, counterarguments: If it is such a great idea, with so many advantages for everyone, why is it not happening more frequently now? Won't the Forest Service be abdicating their responsibilities by pursuing such dispute resolution processes? Aren't environmental and community groups just coopted by such a process?

HOW CAN THE PROCESS BE FINANCED?

Financing can be arranged in a number of different ways. A special line-item appropriation could be requested or special Congressional funding on an experimental basis might be acquired for the program. Probably the most easily accomplished and perhaps the most appropriate manner of financing, however, would be to earmark a certain portion of the oil and gas lease rent and production royalty revenues to a special fund supporting this process. If the process is successful in its objective of reducing the now costly opposition by producing viable decisions at the outset, funding should not actually constitute an addition to the agency's budget. If application review and analysis time and administrative appeal and judicial review costs are reduced by this process, funding should represent an overall *reduction* in agency expenditures.

HOW WILL THE FEDERAL ADVISORY COMMITTEE ACT PROVISIONS BE FULFILLED?

The Federal Advisory Committee Act (FACA) requires that federal officials establishing advisory committees file its "charter, meeting announcements, and minutes in the *Federal Register*; justification statements for the Office of Management and Budget; public announcements of meetings and meeting agendas; invitations to the public to attend meetings and filing of annual reports" (5 U.S.C. app., Sections 1–14). Groups that are subject to this Act's provisions are those "having fixed membership selected by a federal official, created for the purposes of providing advice, having an organizational structure and holding periodic meetings" (5 U.S.C. app., Sections 1–14). Currently, Forest Service officials avoid using advisory committees because they find the Act's requirements to be too cumbersome. Furthermore, Forest Service officials avoid these committees because, as they explain, "selection of committee members can introduce a bias which does not represent the entire range of public interests," and, moreover, not assure the "general public" that it is being represented in decision-making. Another problem noted by the Forest Service is that "membership tends to become permanent because of reluctance to tell members their services are no longer desired. There is danger that deadwood will build up and render the committee ineffective." (U.S. Department of Agriculture, Forest Service, 1973, pp. 59, 61)

As described on pages 235–239, the proposed process goes beyond the purpose of advisory committees as traditionally conceived. Its recommendations are developed more interactively than in the one-way comments of the current advisory committee system. Moreover, its recommendations are then placed before the general public for comment and potential revision. The participants are self-selected with the critical requirement that they be representative of the interests at stake and, furthermore, trusted by their respective constituencies. The Forest Service may want to try to ease fulfillment of FACA's charter requirements by making special arrangements with the Office of Management and Budget for approving participants in selected cases in order to experiment with this process. For example, the Environmental Protection Agency obtained a special charter approval for participants in their experimental rule-making process. Participants involved in the regulatory negotiation processes were considered an advisory group to the EPA and were officially chartered by the Office of Management and Budget's General Services Administration. To comply with FACA, all meetings must be public and meeting schedules must be announced in the *Federal Register* (Susskind and McMahon, 1985, note 69).

WHY DO THESE NEGOTIATIONS NOT OCCUR NOW?

Another hesitation expressed about a dispute settlement approach to decision-making is that, if the advantages are so great, why is it not happening now? As illustrated in Chapters 7 and 8, negotiations occasionally do occur but clearly not as frequently as advocated here. To a large extent, such disputes are not settled more frequently because the established administrative decision-making process does not accommodate it. The institutional structure is not there that would facilitate bargaining and encourage dispute settlement when it might be possible; the needed dialogue is never initiated because the opportunity within which to do so is never provided.

Perhaps a more critical reason negotiations do not occur now is that each group appears to have more power and influence over the final decision outcome through means other than the administrative process. In fact, their power seems least within the established process. And, as seen, because other forums are not designed to acknowledge the legitimacy of all groups and the need for tradeoffs in decision-making, groups have little incentive to help effect the needed tradeoffs. Furthermore, there is little incentive to negotiate in the manner proposed here because there is no assurance that it will make a difference; this approach to decision-making has never been legitimized. The important procedural steps that would encourage collaboration and cooperation have simply never been put into place.

WHAT IS THE ROLE OF THE FOREST SERVICE?

A common Forest Service response to suggested conflict management is that it is the agency's mandate, indeed its *raison d'etre*, to make these decisions. If professional foresters were to let interest group negotiations decide the fate of the national forests, why have a Forest Service at all? Wouldn't agency officials be abdicating their responsibilities?

A counterargument is that it is also the Forest Service's responsibility to represent all values in decision-making and to make decisions in a harmonious manner. There is clearly a need for a scientific, professional Forest Service. Professional expertise and judgment are critical for much of the day-to-day forest management tasks. But, when disputes like those described in this book arise, it is also their responsibility to represent each set of values in their decision. The Forest Service has a critical role in the conflict management process proposed here. The process does not occur outside agency jurisdiction but rather within it, as a supplement to existing administrative procedures. Agency officials need to actively participate in this process both to represent the nonvocal public that is not present as well as to provide the technical and scientific facts and administrative constraints that only they can provide.

IS CONFLICT MANAGEMENT A COOPTING STRATEGY?

The argument has been made that environmental conflict management strategies are designed to coopt environmental and community groups in order to facilitate development (Amy, 1987; Hays, 1987). The process proposed here neither forces participation nor forecloses other avenues if a group believes that the process is not going to serve its best interests. This process provides a choice of *how* to participate; a choice that is lacking in the current process. At the policy level, it allows both industry and environmental organizations to assist in developing broad policies, rather than incrementally questionning policy implementation in numerous site-specific cases. At the site-specific level, it does not preclude the critical precedent-setting nature of many cases. In fact, the agreements reached in site-specific cases, precisely because they will be supported by many different groups, will perhaps have greater precedent-setting value than the cases that now end up receiving conflicting rulings in the courts.

SUMMARY

The conflict-management process proposed here will *not* resolve every national forest management conflict. There will inevitably be cases where precedent-setting issues are of concern that might be better resolved in the courts. There might be cases where agreement between the parties simply cannot be reached and formal administrative appeals and the courts are pursued. There might be cases where procedural or constitutional concerns will need to be reviewed judicially. There are many cases, however, that are resolvable if the institutional structure existed to accommodate and encourage this resolution. Such processes should make the difficult forest management tasks somewhat "easier" and much less adversarial. By developing mutually acceptable outcomes for some forest management decisions, the process will more likely accommodate more public land values than in the win–lose adversarial processes currently applied. Solving part of the problem is certainly better than solving no part at all.

References

Alaska v. Andrus, 580 F.2d 465 (1978).

Altshuler, A. A. *The city planning process: A political analysis*. Ithaca, NY: Cornell University Press, 1965.

Altshuler, A. A., and R. W. Curry. The changing environment of urban development policy—Shared power or shared impotence? *Urban Law Annual*, 1975, *10*(3), pp. 3–41.

Amy, D. J. *The politics of environmental mediation*. New York: Columbia University Press, 1987.

Andrews, R. N. L. Environment and energy: Implications of overloaded agendas. *Natural Resources Journal*, 1979, *19*, 487–503.

Atkin, D. How to monitor, influence and appeal forest service timber sales: A citizen's handbook. Eugene, OR: Oregon Natural Resources Council, 1986.

Bacow, L. The technical and judgmental dimensions of impact assessment. *Environmental Impact Assessment Review*, 1980, *1*(2), 109–124.

Bacow, L., and M. Wheeler. *Environmental dispute resolution*. New York: Plenum Press, 1984.

Bardach, E. *The implementation game: What happens after a bill becomes a law*. Cambridge, MA: MIT Press, 1977.

Barney, D. *The last stand: Ralph Nader's study group report on the national forests*. New York: Grossman, 1974.

Beard, C. A., and M. R. Beard, *The Beards' new basic history of the United States*. New York: Doubleday, 1960.

Behan, R. W. The myth of the omnipotent forester. *Journal of Forestry*, 1966, *64*, 398–407.

Benfield, F. K., J. R. Ward, and R. J. Wilson. *Statement of reasons in support of appeal: In re:appeal of land management plan and environmental impact statement for the San Juan National Forest: Natural Resources Defense Council, Inc., Public Lands Institute, The Wilderness Society, National Audubon Society, Colorado Open Space Council, Colorado Wildlife Federation, Colorado Mountain Club, and San Juan Audubon Society, appellants*. December 5, 1983.

Benfield, F. K., J. R. Ward, and R. J. Wilson. *Appellants' Reply to Responsive Statement: In re:Appeal of Land Management Plan and Environmental Impact Statement for the San Juan National Forest: Natural Resources Defense Council, Inc., Public Lands Institute, The Wilderness Society, National Audubon Society, Colorado Open Space Council, Colorado Wildlife*

249

Federation, Colorado Mountain Club, and San Juan Audubon Society, Appellants. February 23, 1984.

Bingham, G. *Resolving environmental disputes: A decade of experience.* Washington, DC: The Conservation Foundation, 1986.

Bingham, G., B. Vaughn, and W. Gleason. *Environmental conflict resolution: Annotated bibliography.* Washington, DC: The Conservation Foundation, 1981.

Brimmer, C. In the U.S. District Court for the District of Wyoming: Memorandum opinion and order granting plaintiffs motion for summary judgment. *Mountain States Legal Foundation* v. *Andrus,* Case No. C78-165B, October 10, 1980.

Brizee, C. W., Judicial review of Forest Service land management decisions: Part I. *Journal of Forestry,* 1975a, 73(7), 424–425.

Brizee, C. W. Judicial review of Forest Service land management decisions: Part II. *Journal of Forestry,* 1975b, 73(8), 516–519.

Buckle, L. G., and S. R. Thomas-Buckle. Placing environmental mediation in context: Lessons from "failed" mediations. *Environmental Impact Assessment Review,* 1986, 6, 55–70.

Bury, R. and G. Lapotka. The making of wilderness. *Environment,* 1979, 21(10), 12–20.

California v. Bergland, 483 F. Supp. 465 (1980).

California v. Block, 690 F.2d 753 (1982).

Clark, P. B., and W. M. Emrich. *New tools for resolving environmental disputes: Introducing federal agencies to environmental mediation and related techniques.* New York: American Arbitration Association, 1980.

Clawson, M. *The federal lands since 1956: Recent trends in use and management.* Baltimore, MD: Resources for the Future/Johns Hopkins University Press, 1967.

Clawson, M. *The bureau of land management.* New York: Praeger, 1971.

Clean Air Act, 42 U.S.C. 7401 (1977).

Conner v. Burford, 605 F. Supp 107 (1985).

Conservation Law Foundation of New England v. Andrus, 623 F.2d 712 (1979).

Cormick, G. W. Intervention and self-determination in environmental disputes: A mediator's perspective. *Resolve,* Winter, 1982, 1–7.

Coser, L. A., *The functions of social conflict,* New York: The Free Press, 1956.

Coser, L. A., *Continuities in the study of social conflict.* New York: The Free Press, 1967.

Crowell, J., Assistant Secretary of Agriculture for Natural Resources and Environment. *The management of forest lands and natural resources.* Address given to the Symposium on "Public Lands in the Reagan Administration: Access to America's Natural Resources." Co-sponsored by the Government Research Corporation, the Conservation Foundation, the American Mining Congress, and the National Forest Products Association, Denver, Colorado, November 19–20, 1981.

Crowfoot, J. E. Negotiations: An effective tool for citizen organizations? *Northern Rockies Action Group Papers,* 1980, 3(4), 24–44.

Culhane, P. *Public lands politics: Interest group influence and the Forest Service and Bureau of Land Management,* Baltimore, MD: Resources for the Future/Johns Hopkins University Press, 1981.

Cutler, M. R. A study of litigation related to management of Forest Service administered lands and its effect on policy decisions. Unpublished doctoral dissertation, Michigan State University, 1972.

Cutler, M. R. Public involvement in USDA decision-making. *Journal of Soil and Water Conservation,* 1978, 33(6), 264–266.

Dana, S. T., and S. Fairfax, *Forest and range policy* (2nd ed.). New York: McGraw Hill, 1980.

Doherty, W. E. Jr., *Conservation in the United States: A documentary history—Minerals.* New York: Chelsea House/Van Nostrand Reinhold, 1971.

Dravnieks, D., and D. C. Pitcher. *Public participation in resource planning: Selected literature abstracts*, Washington, DC: U.S. Department of Agriculture, 1982.

Edwards, G. C., III. *Implementing public policy*. Washington, DC: Congressional Quarterly Press, 1980.

Edwards, H., Attorney for ANACONDA. *Public lands law and the mining industry*. Address given to the Symposium on Public Lands in the Reagan Administration: Access to America's Natural Resources. Cosponsored by the Government Research Corporation, the Conservation Foundation, the American Mining Congress, and the National Forest Products Association, Denver Colorado, November 19–20, 1981.

Environmental Defense Fund v. Andrus, 596 F.2d 848 (1978).

Endangered Species Act, 16 U.S.C. 1531 (1973).

Evans, B., Vice-President, National Audubon Society. *The environmental community: Responding to the Reagan Administration program*. Address given to the Symposium on "Public Lands in the Reagan Administration: Access to America's Natural Resources." Cosponsored by the Government Research Corporation, the Conservation Foundation, the American Mining Congress and the National Forest Products Association, Denver Colorado, November 19–20, 1981.

Fairfax, S. K. Riding into a different sunset: The sagebrush rebellion. *Journal of Forestry*, 1981, *79*(8), 516–582.

Federal Land Policy and Management Act, 43 U.S.C. 1701(note) (1976).

Federal Water Pollution Control Act, 33 U.S.C. 1324.

Findley, R. Our national forests: Problems in paradise. *National Geographic*, 1982, *162*(3), 306–339.

Fiorino, D. J., and C. Kirtz. Breaking down walls: Negotiated rulemaking at EPA. *Temple Environmental Law and Technical Journal*, 1985, *4*, 29–40.

Fisher, R., and W. Ury. *Getting to yes: Negotiating agreement without giving in*. Boston: Houghton Mifflin, 1981.

Forest and Rangeland Renewable Resources Planning Act, 16 U.S.C. 1601(note) (1974).

Foss, P. *Politics and Grass*. Seattle, WA: University of Washington Press, 1960.

Frank, B. *Our National Forests*. Norman, OK: University of Oklahoma Press, 1955.

Frome, M. *Battle for the wilderness*. New York: Praeger, 1974a.

Frome, M. *The Forest Service*. New York: Praeger, 1974b.

Frome, M. *The Forest Service* (2nd ed.). Boulder, CO: Westview Press, 1984.

Freeman, A. M., III. *The benefits of environmental improvement*. Baltimore, MD: Resources for the Future/Johns Hopkins University Press, 1979.

Gates, P. W. *History of public land law development*, Washington, D.C.: Public Land Law Review Commission, 1968.

Golten, R. J., Mediation: A "Sellout" for conservation advocates, or a bargain? *The Environmental Professional*, 1980, *2*, 62–66.

Goodell, S. K., Waterway preservation: The Wild and Scenic Rivers Act of 1968. *Boston College Environmental Affairs Law Review*, 1978, *7*(1), 43–82.

Gottlieb, K. C. Rigging the wilderness, Pt. I. *The Living Wilderness*, 1982, *45*(156), 13–17.

Hall, G. R. The myth and reality of multiple use forestry. *Natural Resources Journal*, 1963, *3*, 276–290.

Hall, J. F., and R. S. Wasserstrom. The National Forest Management Act of 1976: Out of the courts and back to the forests. *Environmental Law*, *8*, 523–538.

Hanson, D. G. Some plain dealing pays off. *Sierra*, 1987, *72*, 20–22.

Harter, P. J. Negotiating regulations: A cure for the Malaise? *Georgetown Law Journal*, 1982, *71*(1), 1–118.

Harter, P. J. Dispute resolution and administrative law: The history, needs, and future of a complex relationship, *Villanova Law Review*, 1984a, *29*, 1393.

Harter, P. J. Regulatory negotiation: The experience so far. *Resolve,* Fall, 1984b.

Hays, S. P. *Conservation and the gospel of efficiency,* Cambridge, MA: Harvard University Press, 1959.

Hays, S. P. *Beauty, health, and permanence: Environmental politics in the United States, 1955–1985,* Cambridge, England: Cambridge University Press, 1987.

Healey, R. G. *The lands nobody wanted.* Washington, DC: Conservation Foundation, 1977.

Henning, D. H. Natural resources administration and the public interest. *Public Administration Review,* 1970, *30,* 134–140.

Henning, D. H. The ecology of the political/administrative process for wilderness classification. *Natural Resources Journal,* 1971, *11,* 69–75.

Himes, J. S. *Conflict and conflict management.* Athens, GA: University of Georgia Press, 1980.

Horowitz, D. L., *The courts and social policy.* Washington, DC: Brookings, 1977.

Ise, J. *The United States oil policy.* New Haven, CT: Yale University Press, 1926.

Johnson, H. D. The flaws of RARE II. *Sierra,* 1979, *64*(3), 8–10.

Johnson, W. L. Statement of Facts. Civil Action No. 81-1230, *Sierra Club* v. *Peterson,* U.S. District Court for the District of Columbia, 1981.

Kaufman, H. *The forest ranger: A study in administrative behavior.* Baltimore, MD: Johns Hopkins University Press/Resources for the Future, 1960.

Kaufman, H. *The limits of organizational change.* University, AL: University of Alabama Press, 1975.

Kelman, S. Occupational safety and health administration. In J. Q. Wilson, (Ed.), *The politics of regulation.* New York: Basic Books, 1980.

Kuhn, T. S. *The structure of scientific revolutions,* 2nd ed., Chicago: University of Chicago Press, 1970.

Lichfield, N., P. Kettle, and M. Whitbread. *Evaluation in the planning process.* Oxford: Pergamon Press, 1975.

Lowi, T. *The end of liberalism.* New York: W. W. Norton, 1969.

McCleery, D. W. Deputy Assistant Secretary, Department of Agriculture, Memorandum to U.S.Forest Service Chief R. Max Peterson. *USDA decision on review of administrative decision by the Chief of the Forest Service related to the administrative appeals of the forest plans and EISs for the San Juan National Forest and the Grand Mesa, Uncompahgre, and Gunnison National Forest.* Washington, DC: U.S. Department of Agriculture, Office of the Secretary, July 31, 1985.

Managing federal lands: Replacing the multiple use system. *Yale Law Review,* 1973, *82*(4), 787–805.

Martin, J. B. The interrelationships of the Mineral Lands Leasing Act, the Wilderness Act, and the Endangered Species Act: A conflict in search of resolution. *Environmental Law,* 1982, *12,* 363.

McCloskey, M. The Wilderness Act of 1964: Its background and meaning. *Oregon Law Review,* 1966, *45*(4), 288–321.

McConnell, G. *Private Power and American Democracy.* New York: Knopf, 1966.

McNamara, J. Federal Land Management Policy and the drive to develop an alternate energy source, geothermal energy: Shall the twain ever meet? *Natural Resources Journal,* 1979, *19,* 261–274.

Mason, J., and J. Desmond. The land-use fight that didn't happen. *American Forests,* 1983, *89*(11), 17–56.

Mernitz, S. *Mediation of environmental disputes: A sourcebook.* New York: Praeger, 1980.

Merritt, C. R. Shall we lose this heritage? *The Living Wilderness,* 1975, *39*(130), 46–47.

Moore, C. *The mediation process: Practical strategies for managing conflict.* San Francisco: Jossey Bass, 1986.

Mosher, F. C. *Democracy and the public service.* New York: Oxford University Press, 1968.

Mountain States Legal Foundation v. Andrus, 499 F. Supp. 383 (1980).

Multiple-Use Sustained-Yield Act, 16 U.S.C. 528(note) (1960).

Nash, R. *Wilderness and the American mind*, rev. ed. New Haven, CT: Yale University Press, 1967.

National Environmental Policy Act, 42 U.S.C. 4332 (1970).

National Forest Management Act, 16 U.S.C. 1600(note) (1976).

National Materials and Minerals Policy Research and Development Act, 30 U.S.C. 1601 (1980).

National Wildlife Federation v. United States Forest Service, 592 F. Supp. 931 (1984).

Nelson, R. A. Oil and gas leasing on Forest Service lands: A question of NEPA compliance. *The Public Land Law Review*, 1982, 3, 1–50.

Noble, H. Oil and gas leasing on public lands: NEPA gets lost in the shuffle. *Harvard Environmental Law Review*, 1982, 16, 117.

O'Hare, M. Not on my block, you don't—Facility siting and the strategic importance of compensation. *Public Policy*, 1977, 25(4), 407–458.

O'Hare, M., L. Bacow, and D. Sanderson. *Facility siting and public opposition.* New York: Van Nostrand, 1983.

O'Toole, R. Putting your cards on the table: Negotiating a settlement with the Forest Service. *Forest Watch*, 1986, 7(3), 16–19.

Overton, A., American Mining Congress. *Mining and access to public lands.* Address given to the Symposium on "Public Lands in the Reagan Administration: Access to America's Natural Resources. Cosponsored by the Government Research Corporation, the Conservation Foundation, the American Mining Congress, and the National Forest Products Association. Denver, Colorado, November 19–20, 1981.

Pacific Legal Foundation v. James Watt, 529 F. Supp. 982 (1981).

Parker v. United States, 309 F. Supp. 593 (1970).

Perkins v. Bergland, 608 F.2d 803 (1979).

Peffer, E. L. *The Closing of the public domain: Disposal and reservation policies 1900–1950.* Stanford: Stanford University Press, 1951.

Peterson, R. M., Chief, United States Forest Service. Brief of defendants in *Thomas* v. *Peterson.* U.S. Court of Appeals for the Ninth Circuit, On Appeal from the U.S. District Court for the District of Idaho, Ninth Circuit, No. 84-3887, September, 1984.

Pinchot, G. *Breaking new ground.* New York: Harcourt, Brace, and Co., 1947.

Pressman, J. L., and A. B. Wildavsky. *Implementation.* Berkeley, CA:University of California Press, 1973.

Raiffa, H. *The art and science of negotiation.* Cambridge, MA: Harvard University Press, 1982.

Reich, C. *Bureaucracy and the forests.* Santa Barbara, CA: Center for the Study of Democratic Institutions, 1962.

Reilly, W. K., Executive Director, Conservation Foundation. *Public lands: Planning for development and conservation.* Address given to the Symposium on "Public Lands in the Reagan Administration: Access to America's Natural Resources." Cosponsored by the Government Research Corporation, the Conservation Foundation, the American Mining Congress, and the National Forest Products Association. Denver, Colorado, November 19–20, 1981.

Robinson, G. O. *The Forest Service: A study in public land management.* Baltimore, MD: Resources for the Future/Johns Hopkins University Press, 1975.

Robinson, S. D. RARE II: The last act. *Loggers Handbook*, 1979, 39, 24–25, 108–112.

Rocky Mountain Oil and Gas Association v. Andrus, 500 F. Supp. 1338 (1980).

Sabatier, P. A., and D. A. Mazmanian. *Can regulation work?: The implementation of the 1972 California Coastal Initiative.* New York: Plenum Press, 1984.

Schiff, A. *Fire and water.* Cambridge, MA: Harvard University Press, 1962.

Schubert, G. A. Jr. "The public interest" in administrative decision-making: Theorem, theosophy, or theory? *American Political Science Review,* 1957, *51,* 346–368.

Shands, W. E., and R. G. Healy. *The lands nobody wanted.* Washington, D.C.: The Conservation Foundation, 1977.

Sheldon, K. P. *Statement of reasons in support of appeal.* (In the matter of Forest Service recommendations and consent on oil and gas leases issuance involving National Forest System lands in the Palisades further planning area, Idaho and Wyoming,) Sierra Club Legal Defense Fund, 1980a.

Sheldon, K. P. *Reply to the regional forester's responsive statement.* Sierra Club Legal Defense Fund, 1980b.

Shepherd, J. *The forest killers.* New York: Weybright and Talley, 1975.

Sierra Club v. Block, 614 F. Supp. 488 (1985).

Sierra Club v. Hardin, 325 F. Supp. 99 (1971).

Sierra Club v. Hathaway, 579 F.2d 1162 (1978).

Sierra Club v. Peterson 717 F.2d 1409 (1983).

Sirmon, J. M., *Responsive Statement—Palisades further planning area environmental assessment.* Author, 1980(a).

Sirmon, J. M., Regional Forester, Intermountain Region. *Finding of no significant impact—Environmental assessment on Forest Service recommendations and consent for oil and gas lease issuance in the Palisades Further Planning Area.* Author, 1980(b).

Steen, H. K., *The U.S. Forest Service: A history.* Seattle, WA: University of Washington Press, 1976.

Stewart, R. B. The reformation of American administrative law. *Harvard Law Review,* 1975, *88,* 1667–1813.

Stokey, E., and R. Zeckhauser. *A primer for policy analysis.* New York: W. W. Norton, 1978.

Sumner, D. Let's save our roadless areas *The Living Wilderness,* 1977, *40*(136), 25–33.

Sumner, D. Oil and gas leasing in wilderness—What the conflict is about. *Sierra,* 1982, *67,* 28–34.

Stanfield, R. L. Resolving disputes. *National Journal,* 1986, *18*(46), 2764–2768.

Susskind, L., and J. Cruikshank. *Breaking the impasse: Consensual approaches to resolving public disputes.* New York: Basic Books, 1987.

Susskind, L., and G. McMahon. The theory and practice of negotiated rulemaking. *Yale Journal on Regulation,* 1985, *3*(1), 133–165.

Susskind, L., and A. Weinstein. Towards a theory of environmental dispute resolution. *Boston College Environmental Affairs Law Review,* 1981, *9*(2), 311–357.

Susskind, L., J. Richardson, and K. Hildebrand. Resolving environmental disputes: Approaches to intervention, negotiation, and conflict resolution. Unpublished manuscript. MIT Laboratory of Architecture and Planning, June, 1978.

Tableman, M. A. *San Juan National Forest Mediation Case Study.* Unpublished manuscript, University of Michigan, Environmental Conflict Project, School of Natural Resources, Ann Arbor, Michigan, 1985.

Talbot, A. R. *Settling things: Six case studies in environmental mediation.* Washington, D.C.: The Conservation Foundation, 1983.

Tennessee Valley Authority v. Hill, 437 U.S. 153 (1978).

The citizens' guide to timber management in the National Forests. *Forest Planning,* 1985, *6*(7), 1–39.

The Wilderness Society. *How to appeal a forest plan: A citizen handbook,* Washington, DC: Author, 1986a.

The Wilderness Society. *Issues to raise in a forest plan appeal: A citizen handbook.* Washington, DC: Author, 1986b.

The Wilderness Society, Sierra Club, National Audubon Society, Natural Resources Defense Council, and National Wildlife Federation. *National forest planning: A conservationist's handbook* (2nd ed.). Washington, DC: Author, 1983.

Thomas, H., J. Campbell, E. Roberston, Jr., D. J. Grim, G. Waldenmeyer, D. Holland, H. Adkins, E. I. Robertson, O. Groves, J. Badley, P. Resnick, V. Hopfenbeck, E. C. T. Close, F. Wisner, J. Powell, R. Justis, Dixie Outfitters, R. Gormley, L. West, Idaho Conservation League, Idaho Wildlife Federation, Ada County Fish and Game League, D. Cada, E. E. Day, R. L. Day, T. S. Nelson, J. Ward, F. Ward, H. Erickson, R. B. Jones, M. B. Brigham, E. Langworthy, M. Hayes, and J. Zimmer, Jr.. *Brief of appellants in Thomas v. Peterson.* U.S. Court of Appeals for the Ninth Circuit, On Appeal from the U.S. District Court for the District of Idaho, Ninth Circuit, No. 84-3887, August 24, 1984.

Thomas v. Peterson 753 F.2d 754 (1985).

Torrence, J. F. *Responsive statement to the appeal of the San Juan National Forest Land and Resource Management Plan by the Natural Resources Defense Council, Inc., Public Lands Institute, The Wilderness Society, National Audubon Society, Colorado Open Space Council, Colorado Wildlife Federation, Colorado Mountain Club, San Juan Audubon Society.* Appeal, No. 943, U.S. Department of Agriculture, Forest Service, January 23, 1984.

Udall, M. K. Extending the wilderness system. *The Living Wilderness*, 1978, 41(140), 32–34.

U.S. Department of Agriculture, Forest Service. Public involvement and the Forest Service. In, *U.S. Forest Service Administrative Study of Public Involvement Report.* Washington, DC: U.S. Government Printing Office, 1973.

U.S. Department of Agriculture, Forest Service. *RARE II: Final environmental statement, roadless area review and evaluation.* Report No. FS-325. Washington, DC: U.S. Government Printing Office, 1979.

U.S. Department of Agriculture, Forest Service, Targhee and Bridger-Teton National Forests. *Environmental assessment for oil and gas exploration in the Palisades further planning area.* Washington, DC: U.S. Government Printing Office, 1980.

U.S. Department of Agriculture, Forest Service, Rocky Mountain Region, Shoshone National Forest. *Oil and gas exploration and leasing within the Washakie wilderness: Draft environmental impact statement.* Washington, DC: U.S. Government Printing Office, 1981.

U.S. Department of Agriculture, Forest Service, Pacific Southwest Region, San Francisco. *Land management planning direction.* Washington, DC: U.S. Government Printing Office, 1984.

U.S. Department of Agriculture, Forest Service, Cadillac, MI. *Huron-Manistee National Forest land and resource management plan: Draft environmental impact statement.* Washington, DC: U.S. Government Printing Office, 1985.

U.S. Department of Agriculture, Forest Service, Southwestern Region, Albuquerque. *Cibola National Forest land and resource management plan: Proposed amendment.* Washington, DC: U.S. Government Printing Office, 1986.

U.S. Department of the Interior, Bureau of Land Management. *The federal simultaneous oil and gas leasing system: Important new information about the "Government Oil and Gas Lottery".* Washington, DC: U.S. Government Printing Office, 1981a.

U.S. Department of the Interior, Bureau of Land Management. *Oil and gas: Environmental assessment of BLM leasing program* (Dickinson District). Washington, DC: U.S. Government Printing Office, 1981b.

U.S. Department of the Interior, Geological Survey, U.S. Department of Agriculture, Forest Service. *Draft Cache Creek-Bear Thrust environmental impact statement: Proposed oil and gas drilling near Jackson, Teton County, Wyoming* (Bridger-Teton National Forest). Washington, DC: U.S. Government Printing Office, 1981.

U.S. Department of the Interior, Geological Survey, U.S. Department of Agriculture, Forest Service. *Cache Creek-Bear Thrust environmental impact statement: Proposed oil and gas drilling near Jackson, Teton County, Wyoming* (Bridger-Teton National Forest). Washington, DC: U.S. Government Printing Office, 1982.

U.S. Department of the Interior, Geological Survey. Notice to lessees and operators of federal and indian onshore oil and gas leases. Report No. NTL-6, (no date).

U.S. General Accounting Office, *The National Forests—Better planning needed to improve resource management.* Report No. CED-78-133. Washington, DC: U.S. Government Printing Office, 1978.

U.S. General Accounting Office. *Actions needed to increase federal onshore oil and gas exploration and development.* Report to the Congress by the Comptroller General of the United States, No. EMD-81-40. Washington, DC: U.S. Government Printing Office, 1981.

U.S. General Accounting Office. *Accelerated onshore oil and gas leasing may not occur as quickly as anticipated.* Report to the Chairman, Subcommittee on Oversight and Investigations, Committee on Energy and Commerce, House of Representatives. Report No. EMD-82-34. Washington, DC: U.S. Government Printing Office, 1982.

Valfer, E., S. Laner, and D. Dravnieks Moronne. *Public involvement processes and methodologies: An analysis.* Washington, DC: U.S. Department of Agriculture, Forest Service, Management Sciences Staff, 1977.

Wagar, J. A., and W. S. Folkman. Public participation in forest management decisions. *Journal of Forestry,* 1974, *72,* 405–407.

Warm Springs Dam Task Force v. Gribble, 565 F.2d 549 (1977).

Wenner, L. M. *The environmental decade in court.* Bloomington, IN: Indiana University Press, 1982.

Wiebe, R. H., *The search for order: 1877–1920.* New York: Hill and Wang, 1967.

Wiggans, T. Forest users edge towards consensus. *High Country News,* 1983, *15*(21), 6.

Wild and Scenic Rivers Act, 16 U.S.C. 1271(note) (1968).

Wilderness Act, 16 U.S.C. 1121(note) (1960).

Wilkinson, C. F., and H. M. Anderson. Land and resource planning in the National Forests. *Oregon Law Review,* 1985, *64*(1,2), 1–373.

Wilson, J. Q. The politics of regulations. In J. McKie (Ed.), *The social responsibility of business.* Washington, DC: The Brookings Institution, 1973.

Wondolleck, J. Bargaining for the environment: Compensation and negotiation in the energy facility siting process. Unpublished master's thesis, MIT Department of Urban Studies and Planning, 1979.

Wondolleck, J. The importance of process in resolving environmental disputes. *Environmental Impact Assessment Review,* 1985, *5*(4), 341–356.

Wondolleck, J. An evaluation of the U.S. Forest Service natural resource conflict management training program: Final report. Unpublished manuscript, University of Michigan, Environmental Conflict Project, School of Natural Resources, Ann Arbor, 1986.

Wright, H. E. Jr. The boundary waters: Wilderness at stake. *The Living Wilderness,* 1974, *38*(125), 21–31.

Yaffee, S. L. *Prohibitive policy: Implementing the Federal Endangered Species Act.* Cambridge, MA: MIT Press, 1982.

Index